Process Development Symposium 2012

King of Prussia, Pennsylvania, USA
5 – 7 June 2012

ISBN: 978-1-5108-1593-3

Printed from e-media with permission by:

Curran Associates, Inc.
57 Morehouse Lane
Red Hook, NY 12571

Some format issues inherent in the e-media version may also appear in this print version.

Printed by Curran Associates, Inc. (2015)

AIChE
120 Wall Street, FL 23
New York, NY 10005-4020

Phone: (800) 242-4363
Fax: (203) 775-5177

www.aiche.org

Additional copies of this publication are available from:

Curran Associates, Inc.
57 Morehouse Lane
Red Hook, NY 12571 USA
Phone: 845-758-0400
Fax: 845-758-2634
Email: curran@proceedings.com
Web: www.proceedings.com

TABLE OF CONTENTS

Addressing Pharmaceutical Process Development Challenges

Futran, M., Rutgers University

Originally presented on: 6/6/2012 9:10:00 - 9:45:00

The Pharmaceutical Industry is undergoing major changes. More than $100 Billion in patent protected products are facing generic competition this decade. One of the most important changes in "Big Pharma" is an evolution from the vertically integrated model to a distributed one, where innovation, project execution and manufacturing are outsourced, often to Asia. Pharmaceutical process development and manufacturing are included in this trend. The new Pharmaceutical supply chain spans the globe, and involves companies with varying degrees of technical sophistication and operational discipline. Ensuring the supply and quality of pharmaceutical products requires a renewed emphasis on acquiring fundamental understanding during development so that the process can be controlled effectively in this array of suppliers. Furthermore the time and cost of development must be reduced.

This environment represents an opportunity for the academic community to contribute to the science and technology needed in process development and manufacturing, partnering with Industry along the way. While the FDA has created their "Quality by Design" initiative, it is not possible for every company around the world to develop this independently. The talk presents examples of current and potential academic contributions, such as dynamic analysis for process control, the science of particulate systems and the development of continuous manufacturing for solid oral dosage forms, and perspectives for the use of continuous approaches in developing bio-manufactured products.

Investigation of the Impact of Mode of Operation (Batch vs Flow) on Design Space Map

Vaidyaraman, S., Eli Lilly and Company
Originally presented on: 6/6/2012 9:45:00 - 10:15:00

This talk will discuss how the mode of operation (batch vs flow) impacts the design space map for impurities thus impacting the robustness of the process. A specific case study involving a reaction model will be used for this investigation that builds upon previous work on failure mode analysis in continuous processes (Wong etal 2011). There are three impurities of interest formed in this reaction. A reaction kinetic model developed from batch experiments is used to design for both batch and flow processes. The key questions that this investigation addresses are:

- How much does the mode of operation impact the size of the design space?
- Is the link between the uncertainties due to routine variability and model uncertainty and the output impurity levels same or different depending on the mode of operations?
- Can certain generalizations be drawn based on the results of this case study?

Risk Management for New Process Technology Implementation – Some Key Components

Mike Schaller and Aaron Sarafinas
The Dow Chemical Company

AIChE 2012 Process Development Symposium, King of Prussia, PA
Session I: Robust Process and Product Design: Accommodating Uncertainty
June 6, 2012

The Dow Chemical Company

- Founded in 1897 by Herbert H. Dow in Midland, Michigan
- Supplies a broad range of products and services to customers in ~160 countries
- Manufactures more than 5,000 products at 197 sites in 36 countries
- Employs 52,000 employees worldwide
- $60 billion in sales (2011)
- 2009 Acquisition of Rohm and Haas makes Dow the leading specialties chemicals and advanced materials company

- Our Vision
  - To be the most profitable and respected science-driven chemical company in the world
- Our Mission
  - To passionately innovate what is essential to human progress by providing sustainable solutions to our customers.

An Integrated Portfolio Designed for Growth

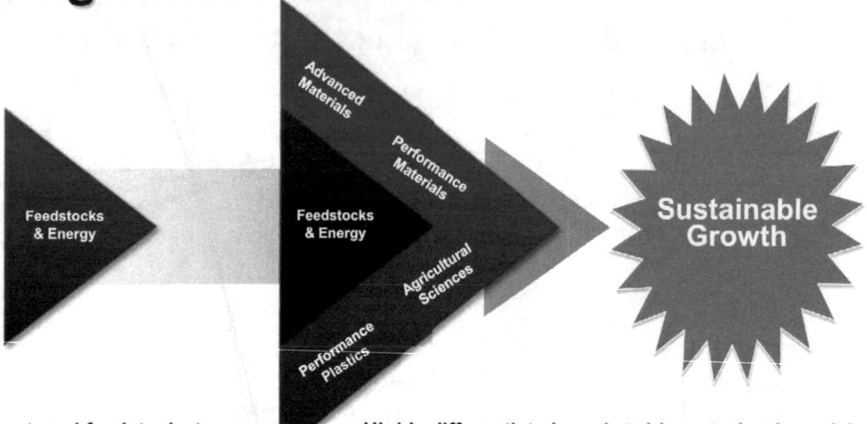

- **Advantaged feedstocks to support and advance downstream growth**
- **Highly differentiated, market-driven, technology-rich businesses focused on addressing megatrends and key end-markets**
  - **Packaging**
  - **Electronics**
  - **Hygiene & Medical**
  - **Water**
  - **Coatings**
  - **Agriculture & Nutrition**
  - **Oil & Gas**

Preview of Presentation

- Premise
  - Capital projects that implement new process technology face extraordinary risk compared to those implementing demonstrated technology.
  - That extraordinary risk can and should be met with extraordinary Risk Management activities.
- Outline
  - Risks facing capital projects with new process technology
  - Background of our internal efforts to improve our performance
  - Risk Management definitions
  - Why Risk Management is difficult
  - Detailed discussion of key Risk Management components
    - Develop and implement a formal Risk Management Plan
    - Incorporate inherently low-risk approaches
    - Incorporate technology development into the critical path project plan
    - Practice thorough, effective risk assessment
    - Prepare for problems

Capital projects with new technology are more challenging

- All capital projects face risk, because we cannot predict with certainty how the project will proceed
  - Scope clarity and scope changes
  - Resource availability and quality
  - Raw material quality and availability
  - Construction labor rates and availability
  - Weather…
- Capital projects with new technology carry all of the possible capital project risks, plus additional ones
  - Compressed schedules
  - Concurrent product and process design
  - More and more frequent scope changes
  - Overlap between process development and process design…
- Initial anecdotal evidence was supported with project data
  - The presence of new process technology adversely affects capital cost, performance against schedule, and operability

New Process Technology Implementation Team

Opportunity: Foster a more collaborative environment between R&D, Manufacturing and Engineering (M&E) to successfully Develop, Transfer, Implement and Start-Up New Process Technology (NPT)

One portion of the larger NPT Implementation effort is to make Risk Management activities more:

- **Formal**
- **Deliberate**
- **Visible**
- **Effective**

What is Risk Management?

- Long Definition*
  - "The identification, assessment and prioritization of risks, followed by coordinated and economical application of resources to minimize, monitor and control the probability and/or impact of unfortunate events."

* Douglas W. Hubbard, The Failure of Risk Management: Why It's Broken and How to Fix It, Wiley, April 27, 2009

What is Risk Management?

- Long Definition*
  - "The identification, assessment and prioritization of risks, followed by coordinated and economical application of resources to minimize, monitor and control the probability and/or impact of unfortunate events."
- Short Definition*
  - "Being smart about taking chances."

* Douglas W. Hubbard, The Failure of Risk Management: Why It's Broken and How to Fix It, Wiley, April 27, 2009

Human Nature Makes Risk Management Challenging

- Source
  - *Drilling for Certainty* , By DAVID BROOKS
  - Published: May 27, 2010 in The New York Times
- Brooks' theme: natural tendencies make people inherently ill-suited to dealing with risk
  - "People have trouble imagining how small failings can combine to lead to catastrophic disasters," especially when the issues are complex.
  - "People have a tendency to get acclimated to risk."
  - "People have a tendency to place elaborate faith in backup systems and safety devices."
  - "People have a tendency to match complicated technical systems with complicated governing structures."
  - "People tend to spread good news and hide bad news."
  - "People in the same field begin to think alike, whether they are in oversight roles or not."
- Brooks does does not imply that people are incapable of dealing with risk
  - Merely, but critically, that we need to make *conscious effort* to avoid our natural, irrational tendencies, lest we fall prey to them

Link to column: http://www.nytimes.com/2010/05/28/opinion/28brooks.html

Our story so far...

- Capital projects with new process technology face extraordinary risk
  - Our premise is that extraordinary risk must be met with extraordinary risk management
  - People's natural tendencies make them ill-suited to assessing and managing risk
- How should we respond?
  - By making our risk management practices more formal, deliberate, visible, and effective
- What are some specific actions we can take?
  - Develop and implement a formal Risk Management Plan
  - Incorporate inherently low-risk approaches
  - Incorporate technology development into the critical path project plan
  - Practice thorough, effective risk assessment
  - Prepare for problems

A Risk Management Process

Keys to success

- Identify a Risk Manager
- Provide support and training
- Perform the process iteratively
- Communicate
- Communicate
- Communicate

Use Low-Risk Approaches in Process Design

- Some of the most important
  - Favor batch processing instead of continuous
  - Eliminate or minimize recycle streams
  - Utilize partnerships and/or contract manufacturing where appropriate
- These and other approaches can be incorporated into a Multi-Generational Plan
  - Limit the scope wherever possible
  - Three singles score as many runs as a solo homer

Why Favor Batch Processing Over Continuous?

- Batch processing is not always less risky than continuous, but if used properly, it can offer significant advantages in implementing new technology
  - Reducing initial cost, while providing easy capacity increase capability
  - Trading rate issues for time issues

Batch Processing Advantages

Capital Cost Minimization / Capacity Increase Capability

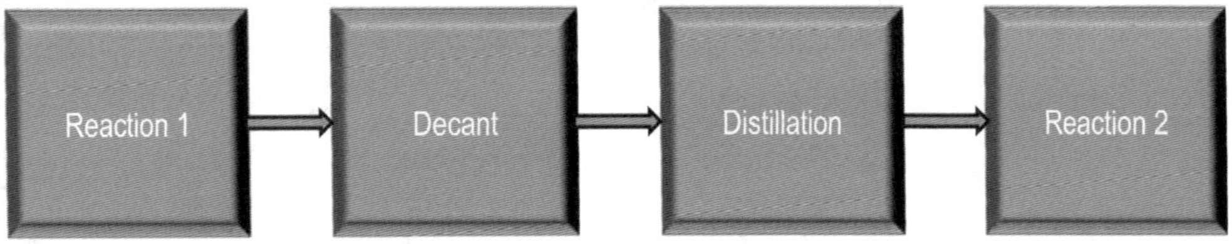

- ID most likely rate-limiting-step
- Combine multiple non-rate-limiting-steps within single vessels
- Add additional vessels as capacity demands
  - And, by the way, after developing commercial-scale performance data

Batch Processing Advantages, *continued*

Trade Rate Issues for Time Issues

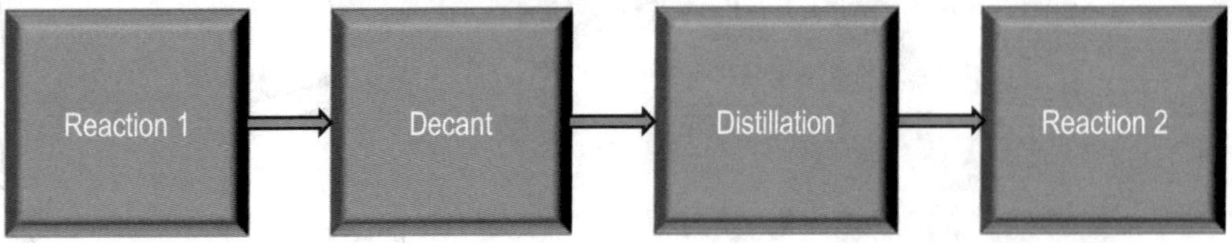

- We often lack the data we need to precisely size continuous equipment or determine cycle time for batch equipment
- In a continuous process, turn-down capability of different units may be different
  - Example: you have to run your decanter at half rate to make the phase separation, but your distillation column control suffers tremendously as a result
- You still pay the overall capacity penalty in a batch process, but that penalty is less likely to cause additional problems

Detailed Technology Development Planning

- The *critical path project plan* is a model of how we will execute the project
  - Its use encourages (demands?) quantitative thinking
- The project plan is critical to risk management
  - It is a tool for identifying and assessing risks
- To be effective, the plan must include details for key technology development activities
  - Accomplishment-based
  - Clearly-defined outcomes based on process and equipment design requirements
- It can provide "insulation" against unrealistic demands
  - Unrealistic demands lead to inappropriate pressure and poor decisions
- The need for invention and discovery, and their inherent uncertainty, does NOT preclude the use of a detailed project plan
  - "All models are wrong, but some are useful" – G.E. Box

Thorough, Effective Risk Assessment

- Make it model-based
- Make it quantitative
- Include perspectives from outside the team
  - People working together on the same effort tend to develop a similar perspective
  - You can't put your best people on every project
    - But you can give many of them a peek or two at many projects

Model-Based, Quantitative Risk Assessment is Difficult

- Why do we expend the time and effort?
  - Because the risk quantification:
    - Increases the rigor of our thought processes
    - Helps identify effective, efficient risk response opportunities
  - Because the model summarizes the risk
    - Risk quantification typically requires risk decomposition
      - The model helps us see the high-level effects of the decomposed risks

Quantitative Assessment

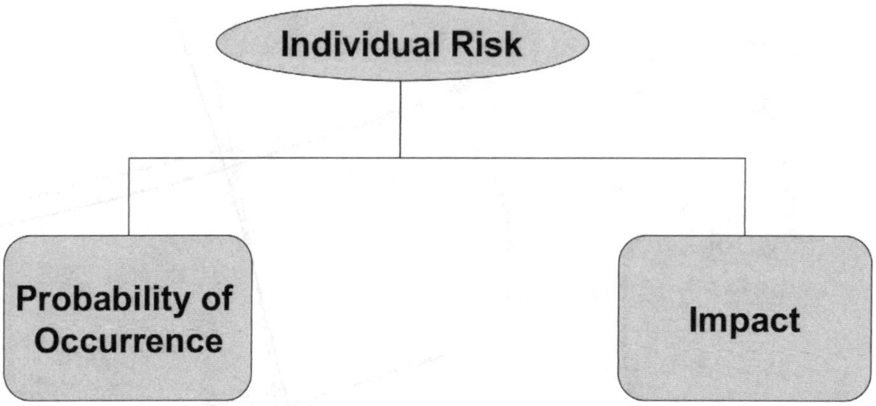

Example: Schedule Risk Model

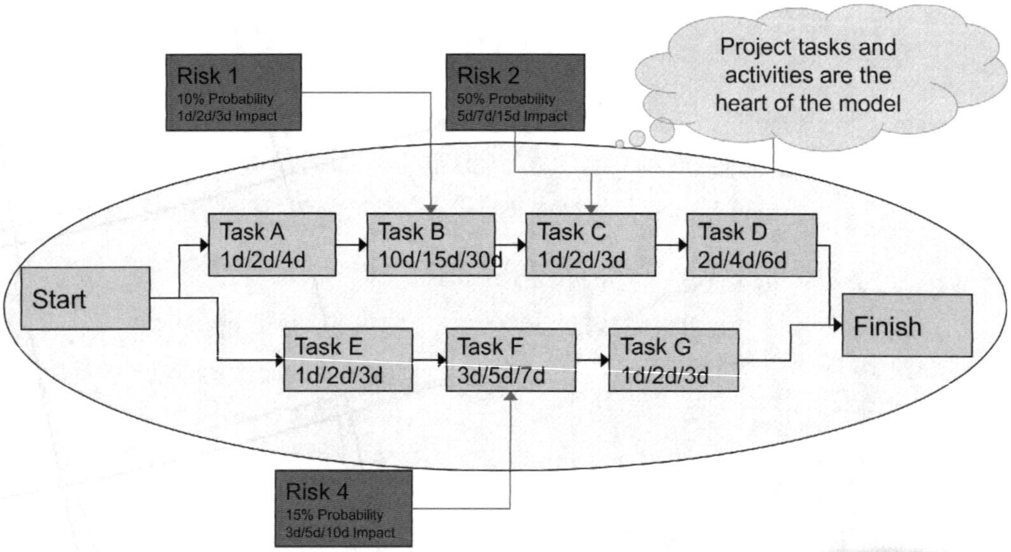

Example: Schedule Risk Model, *continued*

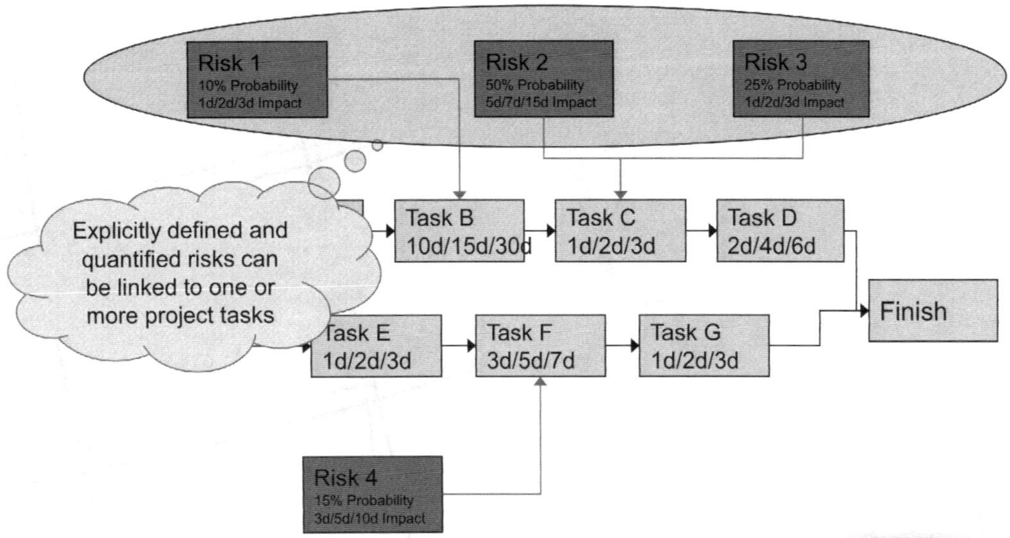

MES 16-May-12 AIChE Process Development Symposium, Session I 21

Schedule Risk Model, *continued*

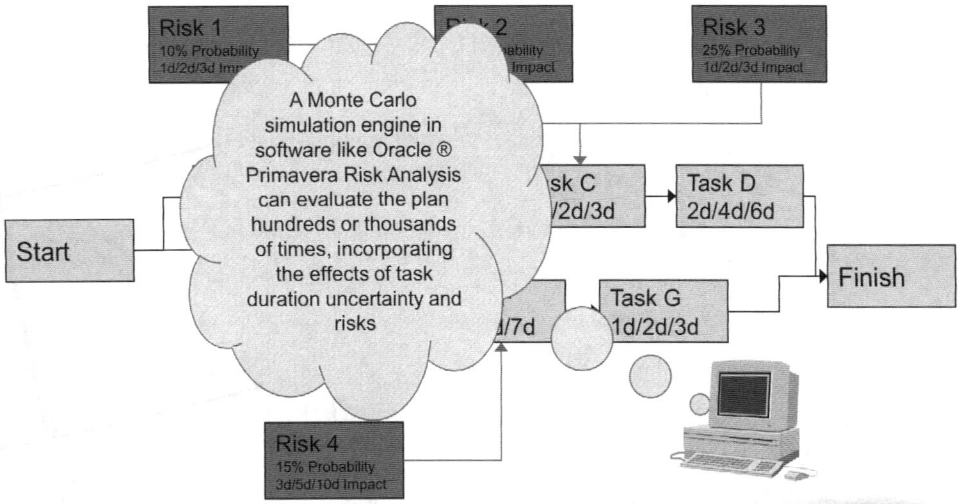

MES 16-May-12 AIChE Process Development Symposium, Session I 22

Schedule Risk Model Results

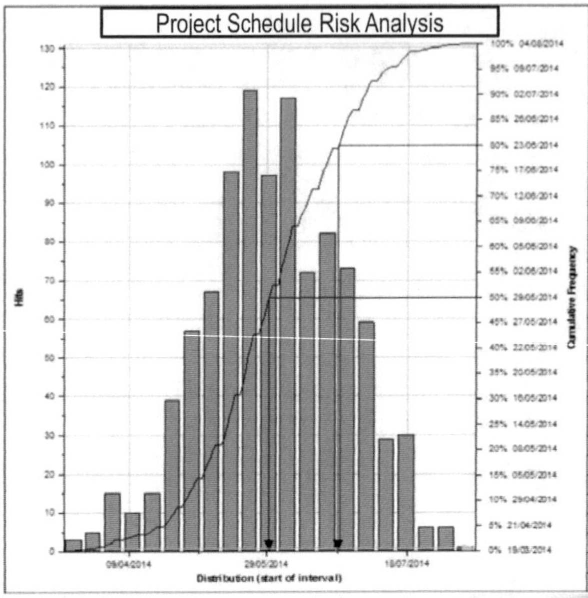

Data	
Finish Date of	
Entire Plan	
Analysis	
Iterations.	1000
Statistics	
Minimum	19/03/2014
Maximum.	04/08/2014
Mean	30/05/2014
Bar Width.	week
Highlighters	
Deterministic (28/03/2014)	<1%
50%	29/05/2014
80%	23/06/2014

Prepare for Problems – Why?

- We cannot predict all potential problems
- We do not have the resources to completely address all those we *can* predict
  - And we probably shouldn't spend too much time fixing *all* the things that *might* go wrong
- "Active acceptance" can be a very effective risk response

=> Preparation for effective problem solving should be a component of the risk management plan for projects involving new technology

Prepare for Problems – How?

- Design the process to produce *information* as well as product
- Troubleshooting Guides
  - A Failure Modes and Effects Analysis (FMEA) and other risk assessments can help generate these
- Characteristics Matrix
  - Maps key product performance parameters to the portions of the process that can affect them, and
  - Identifies where potential root causes "accumulate", making problem solving challenging
  - Identifies opportunities for improved error detection capability
- Plans to provide analytical or other error-detection capability
  - This is typically above and beyond that which is required for the "run plant normal" mode of operation
- Start-up resource plan – people and training
  - Identify the types of skill sets you might need, and prepare them in advance, so that they can "hit the ground running" when needed

Summary

- Implementing capital projects that involve new process technology is challenging
- One part of our effort to improve our NPT implementation performance is to meet increased risk with atypically effective risk management
- Human tendencies hinder effective risk assessment and management
- To counter these tendencies, and to be effective, our risk management efforts should be
  - Formal
  - Deliberate
  - Visible
- Some activities we think especially effective
  - Develop and implement a formal Risk Management Plan
  - Incorporate inherently low-risk approaches
  - Incorporate technology development into the critical path project plan
  - Practice thorough, effective risk assessment
  - Prepare for problems

Recommended Reading

- Irrational Tendencies
  - Thinking, Fast and Slow, by Daniel Kahneman, Farrar, Straus and Giroux, 2011
  - Stumbling on Happiness, by Daniel Gilbert, Vintage Books, 2007
- Risk Assessment and Risk Management
  - How to Measure Anything: Finding the Value of Intangibles in Business, by Douglas W. Hubbard, Wiley, 2010 (2nd edition)
  - The Failure of Risk Management: Why It's Broken and How to Fix It, by Douglas W. Hubbard, Wiley, 2009
- Probabilistic Thinking
  - Why Can't You Just Give Me The Number: An Executive's Guide to Using Probabilistic Thinking to Manage Risk and to Make Better Decisions, by Patrick Leach, Probabilistic Publishing, 2006

Biopharmaceutical Purification Process Development: Identifying Risks and Dealing with Uncertainty

Weisser S. M., GlaxoSmithKline

Originally presented on: 6/6/2012 10:45:00 - 11:15:00

Process and product knowledge is gained throughout the lifecycle of a biopharmaceutical product. Process development teams need to use purification strategies that allow them to respond to new information and accommodate uncertainty. Development of biopharmaceutical purification processes will be discussed with an emphasis on how project teams deal with uncertainty. Strategies focusing on identification of risks and knowledge gaps, and designing and implementing mitigation plans to deal with them will be discussed. Application of these approaches to problems such as dealing with feed stream variability, risks to key raw materials, and issues observed after scale-up will be presented.

PDS Process Development Symposium
June 5-7 2012
King of Prussia, PA

THANKS TO OUR SPONSORS!

ProSim
Software & Services In Process Simulation

DOW

PSE

Scale-up Systems

EASTMAN

Chemstations™

PDS Process Development Symposium
June 5-7 2012
King of Prussia, PA

Introduction to Inherently Safer Design

Dennis C. Hendershot
CCPS Staff Consultant
dennis.hendershot@gmail.com

Dennis Hendershot

Dennis retired in 2005 as a Senior Technical Fellow in the Risk Management Services Department of the Rohm and Haas Company Engineering Division. He worked at Rohm and Haas for 35 years, starting in process research and development, plant design, and startup. From 1980, he focused on process safety and risk management, following an assignment to participate in a pilot evaluation of HAZOP. He worked as a consultant for Chilworth Technology from 2005-2008, and has been a CCPS Staff Consultant since 2005. From 2005-2007 Dennis was a member of the BP North American Refineries Independent Safety Panel ("the Baker Panel") which investigated safety culture issues related to the 2005 Texas City explosion. Dennis has a BS from Lehigh University and an MS from the University of Pennsylvania, both in Chemical Engineering.

CCPS projects include:

Inherently Safer Design, 2nd Edition (2009)
Guidelines for Developing Quantitative Safety Risk Criteria (2009)
Process Safety Beacon
Process Safety Moments
Process Safety Boot Camp Training Program

Dennis has been involved in more than 10 CCPS books and other projects going back to 1987.

Introduction to Inherently Safer Design

Dennis C. Hendershot
CCPS Staff Consultant
dennis.hendershot@gmail.com

Inherently safer design focus

- Safety – immediate impacts of single events
  - People
  - Environment
  - Property and business – "Loss Prevention"
- Fires, explosions, immediate toxic impacts
- Most of these events will also have long term health and environmental impacts

5

History of inherently safer design concept

- Technologists have always tried to eliminate hazards
  - Some examples:
    - In-situ manufacture of nitroglycerine in 1860s railroad construction
    - Alfred Nobel – dynamite in place of pure nitroglycerine for mining, construction
- Trevor Kletz, ICI, UK (1977)
  - Jubilee Lecture to the Society of the Chemical Industries, Widnes, England
  - Response to 1974 Flixborough, UK explosion
  - Named the concept
  - Developed a set of design principles for the chemical industry

6

What is inherently safer design?

- Inherent - "existing in something as a permanent and inseparable element..."
  - safety "built in", not "added on"
- Eliminate or minimize hazards rather than control hazards
- Potential benefit – simpler, cheaper, safer plants
- More a philosophy and way of thinking than a specific set of tools and methods

7

Hazard

- An inherent physical or chemical characteristic that has the potential for causing harm to people, the environment, or property (CCPS, 1992).
- Hazards are intrinsic to a material, or its conditions of use.
- Examples
  - Chlorine - toxic by inhalation
  - Gasoline - flammable
  - High pressure steam - potential energy due to pressure, high temperature

8

Chemical Process Safety Strategies

9

Inherent

- Eliminate or reduce the hazard by changing the process or materials to use materials or conditions which are non-hazardous or less hazardous
- Integral to the product, process, or plant - cannot be easily defeated or changed without fundamentally altering the process or plant design
- EXAMPLE
  - Substituting water for a flammable solvent (latex paints compared to oil base paints)

10

Passive

- Minimize hazard using process or equipment design features which reduce frequency or consequence without the active functioning of any device
- EXAMPLE
  - Conducting a chemical reaction capable of generating a maximum of 5 bar pressure in a reactor designed for 10 bar

11

Active

- Controls, safety instrumented systems (SIS)
- Multiple active elements
  - Sensor - detect hazardous condition
  - Logic device – receive signal from sensor, decide what to do, send signal to control element
  - Control element - implement action
- Prevent incidents, or mitigate the consequences of incidents
- EXAMPLES
  - High level alarm in a tank shuts the feed valve
  - Fire protection – sprinkler system

12

Procedural

- Standard operating procedures, safety rules and standard procedures, emergency response procedures, training

- EXAMPLE
  - An operator is trained to observe the temperature of a reactor and apply emergency cooling if it exceeds a specified value

13

Which strategy should we use?

- Generally, in order of robustness and reliability:
  - Inherent
  - Passive
  - Active
  - Procedural
- But you will need all of them – especially when considering the multiple hazards in any chemical process or product
- Inherent strategies often involve changes to basic process chemistry and unit operations – best considered as early in process development as possible.
- But – it is never too late for inherently safer design!

14

No clear boundaries!

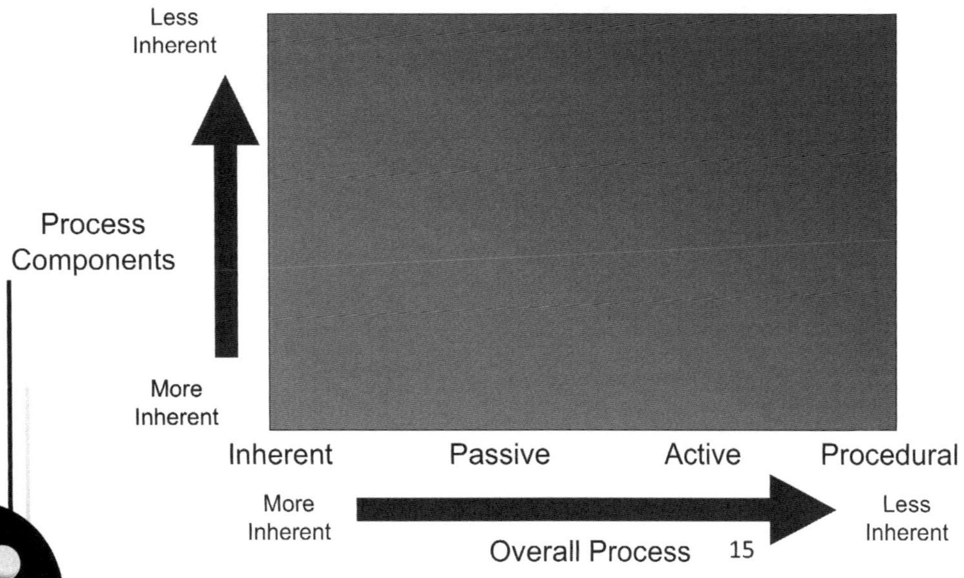

15

ISD and risk

- Risk = f (consequence, likelihood)
- Reduce risk by reducing or eliminating the hazard
  - Eliminate or reduce potential consequence part of the risk equation
- Reduce risk by reducing the ability of the system to allow a potential accident scenario to occur
  - Reduce the frequency part of the risk equation
  - Reduce frequency by equipment features which are a part of the process itself
  - Not "inherent" if it involves add-on safety devices

Event tree model

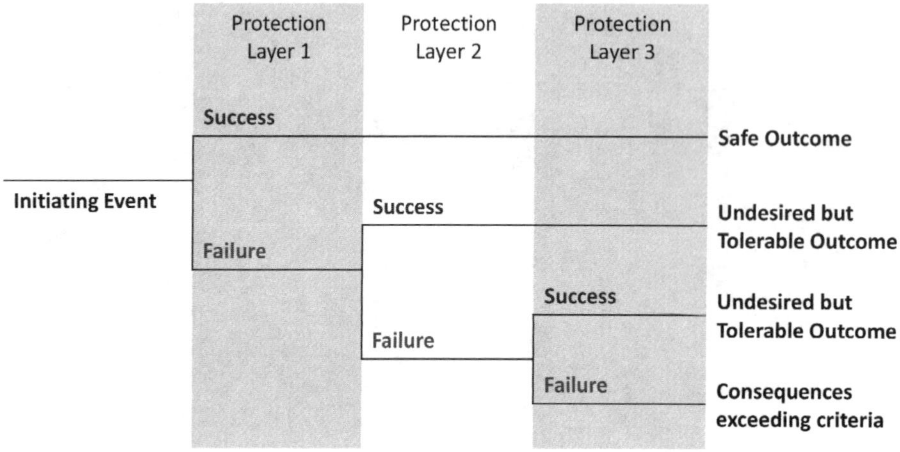

17

ISD as a consequence reduction tool

1. Initial Situation

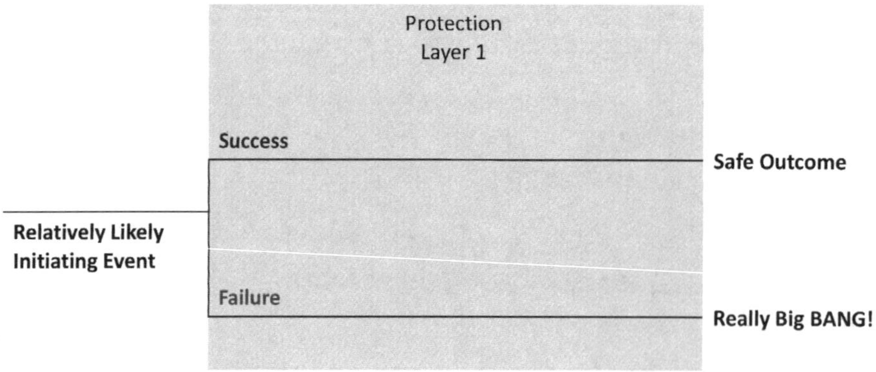

18

ISD as a consequence reduction tool

1a. Traditional Approach – Add PL to meet Risk Criteria

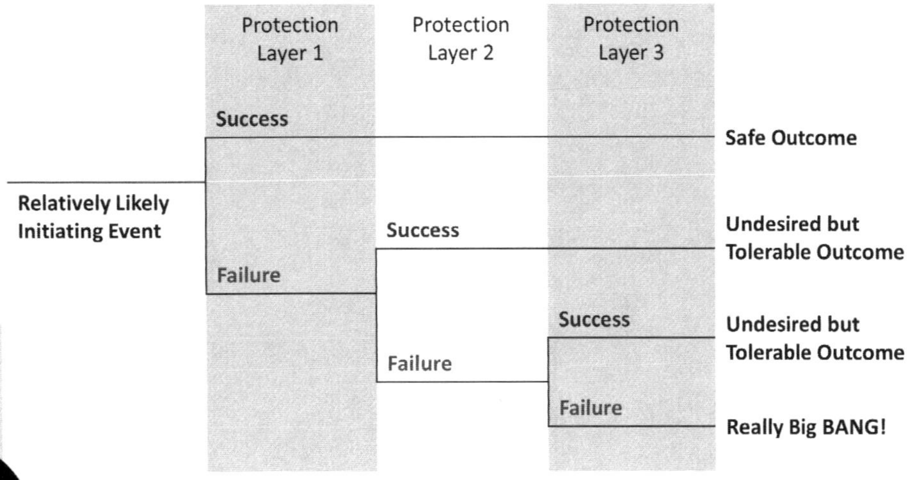

19

ISD as a consequence reduction tool

1b. ISD Approach – Reduce Consequence

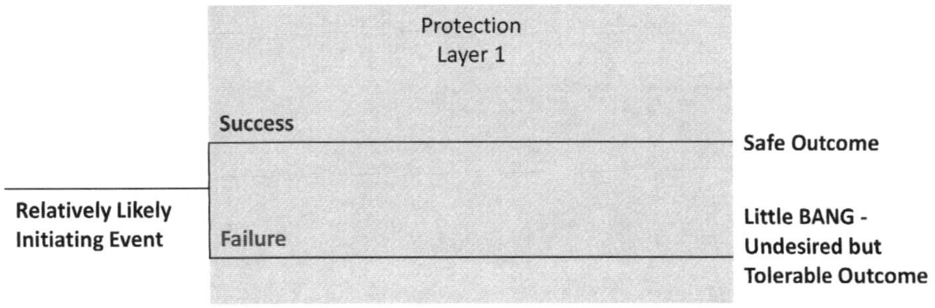

Or, even Inherently Safer (no Protection Layers needed)

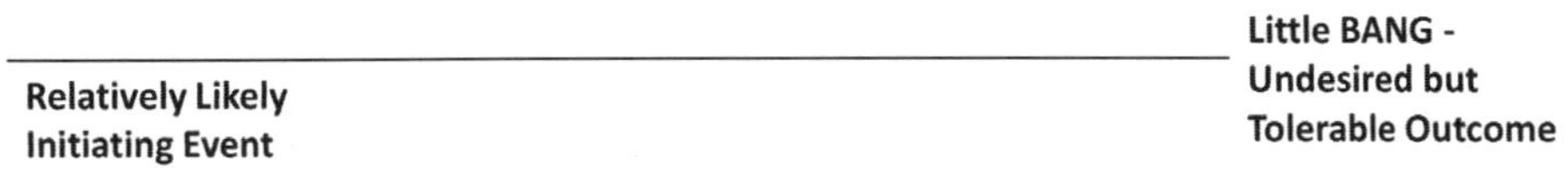

20

ISD as a frequency reduction tool

2. Initial Situation

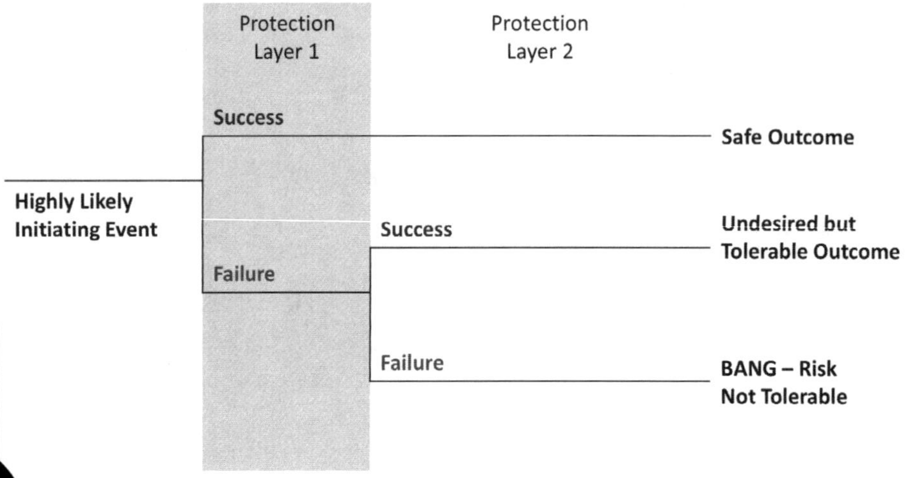

21

ISD as a frequency reduction tool

2a. Traditional Approach – Add LOP to meet Risk Criteria

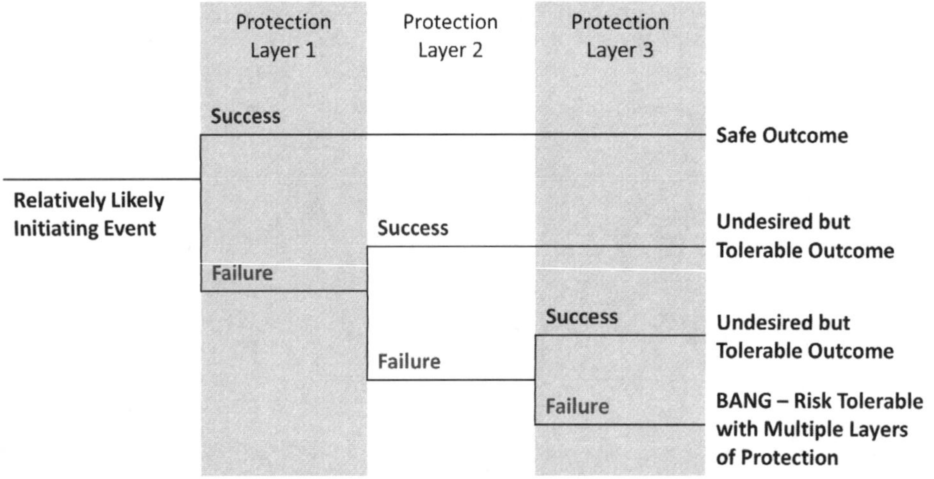

22

ISD as a frequency reduction tool

2b. ISD Approach – Initiating Event is Difficult/Unlikely (or, best of all, IMPOSSIBLE!)

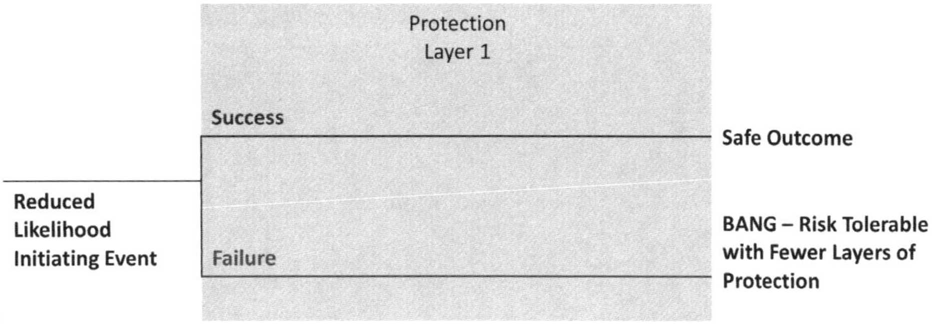

23

Inherently safer design strategies

- Substitute
- Minimize
- Moderate
- Simplify

24

Substitute

- Substitute a less hazardous reaction chemistry
- Replace a hazardous material with a less hazardous alternative

25

Reaction Chemistry - Acrylic Esters

- Reppe Process

$$CH \equiv CH + CO + ROH \xrightarrow[HCl]{Ni(CO)_4} CH_2 = CHCO_2R$$

- Acetylene - flammable, reactive
- Carbon monoxide - toxic, flammable
- Nickel carbonyl - toxic, environmental hazard (heavy metals), carcinogenic
- Anhydrous HCl - toxic, corrosive
- Product - a monomer with reactivity (polymerization) hazards

26

Alternate chemistry

– Propylene Oxidation Process

$$CH_2 = CHCH_3 + \frac{3}{2}O_2 \xrightarrow{Catalyst} CH_2 = CHCO_2H + H_2O$$

$$CH_2 = CHCO_2H + ROH \xrightarrow{H^+} CH_2 = CHCO_2R + H_2O$$

- Inherently safe?
- No, but inherently *safer*. Hazards are primarily flammability, corrosivity from sulfuric acid catalyst for the esterification step, small amounts of acrolein as a transient intermediate in the oxidation step, reactivity hazard for the monomer product.

27

By-products and side reactions

- Organic intermediate production
  - Intended reaction - hydrolysis done in ethylene dichloride solvent

    Organic raw material + sodium hydroxide --->
    product + sodium salt
- Reaction done in ethylene dichloride solvent

28

Hazardous side reaction

- Sodium hydroxide + ethylene dichloride solvent:

$$C_2H_4Cl_2 + NaOH \quad -\succ C_2H_3Cl + NaCl + H_2O$$

- The product of this reaction is vinyl chloride (health hazard)
- A different solvent (perchloroethylene) was used

29

Substitute less hazardous materials

- Organic solvents with a higher flash point and/or lower toxicity for
  - Paints and coatings
  - Dyes
  - Agricultural product formulations
  - Dibasic ethers and organic esters as paint removers
- Aqueous emulsions

30

Minimize: Example – A Nitration Process

- **UNDERSTAND YOUR PROCESS, WHATEVER IT IS!**
- This is the key to identifying inherently safer design options.
- What are the important process parameters that impact the design of your process equipment?

$$\text{Organic substrate (X-H)} + HNO_3 \xrightarrow[\text{Solvent}]{H_2SO_4} \text{Nitrated Product (X-NO}_2\text{)} + H_2O$$

31

Understand your process!

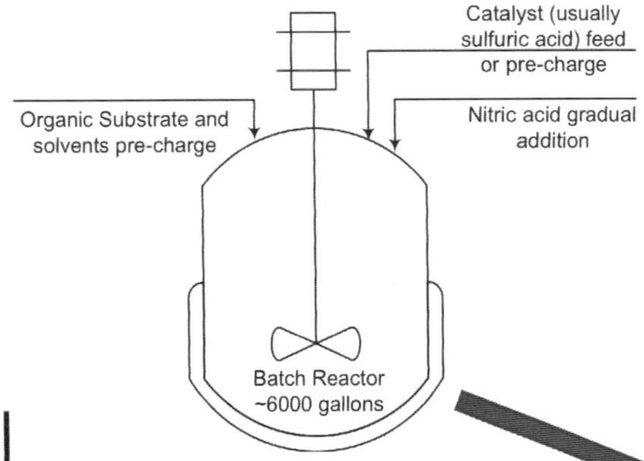

6000 gallon batch reactor

100 gallon continuous reactor

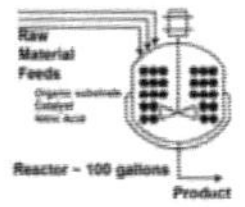

Why is the smaller reactor safer?

32

Storage and Transfer Examples

- General principals
  - Storage of hazardous raw materials should be minimized
    - But - consider the conflicting hazards
      - Transportation hazards
      - Potential increased frequency of plant shutdown
  - Pipes should be large enough to do the required job , and *no larger*
  - Intermediate storage - is it really needed?

33

Moderate

- Dilution
  - Aqueous ammonia instead of anhydrous
  - Dynamite instead of nitroglycerine
- Refrigeration
  - Chlorine storage
  - Other liquefied gases (methylamines, ammonia, etc.)
- Less severe operating conditions (lower temperature and pressure)
  - Improved catalysts
  - Alternate chemistry

34

Simplify

- Eliminate unnecessary complexity to reduce risk of human error
  - *<u>QUESTION ALL COMPLEXITY!</u>* Is it really necessary?

35

Controls on a stove

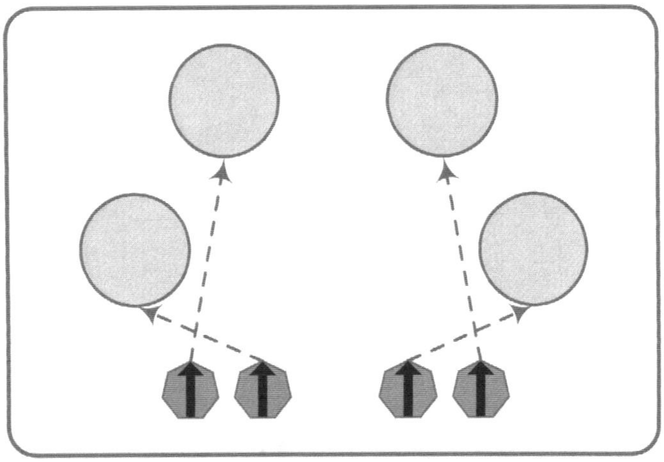

From Don Norman, "Turn Signals are the Facial Expressions of Automobiles"

36

Traditional Methyl Acetate Process

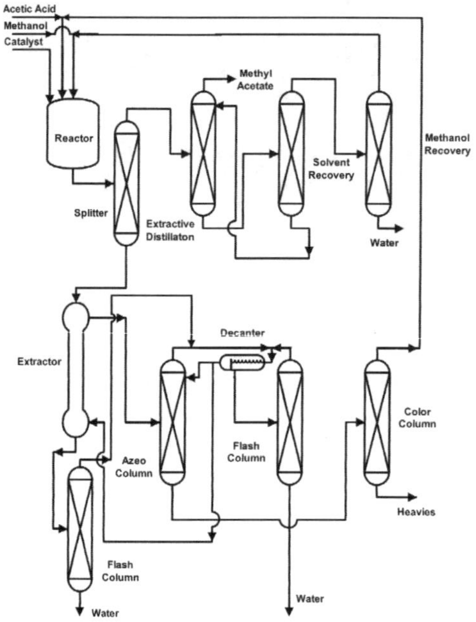

Reactive Distillation Process - Eastman Chemical

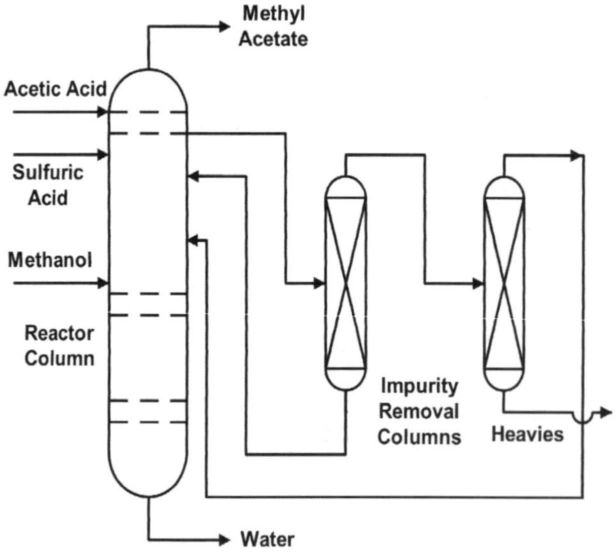

Inherent safety at various levels of process design

- Overall technology
  - What technology for drinking water treatment (disinfection) - chlorine, ozone, UV, others?
- Implementation of the selected technology
  - How is water chlorination to be implemented (chlorine gas, sodium hypochlorite, other ways of chlorinating water)

39

Inherent safety at various levels of process design

- Detailed design for selected technology
  - Water treatment - size of equipment, operating conditions, general layout of plant, single large system or multiple smaller systems, etc.
- Detailed equipment design
  - Water treatment - selection of specific pieces of equipment, location of equipment and piping, location of valves, controls, etc.
- Operation
  - User friendly operating procedures, maintenance procedures, etc.

40

Some problems

- The properties of a technology which make it *hazardous* may be the same as the properties which make it *useful*
  - Airplanes travel at 600 mph
  - Gasoline is flammable
  - Chlorine is toxic
- *Control* of the hazard is the critical issue in safely getting the benefits of the technology

41

Multiple hazards

- Everything has multiple hazards
  - Automobile travel
    - velocity (energy), flammable fuel, exhaust gas toxicity, hot surfaces, pressurized cooling system, electricity......
  - Chemical process or product
    - acute toxicity, flammability, corrosiveness, chronic toxicity, various environmental impacts, reactivity.......

42

What does inherently safer mean?

- Inherently safer is in the context of one or more of the multiple hazards
- There may be conflicts
  - Example - CFC refrigerants
    - Low acute toxicity, not flammable
    - Environmental damage, long term health impacts
    - Are they inherently safer than alternatives such as propane (flammable) or ammonia (flammable and toxic)?
    - "Green" refrigerators available in Europe – use ~ 100 grams hydrocarbon, but required a significant re-design to minimize flammable material inventory.

43

Different Concerns

- Different populations may perceive the inherent safety of different technology options differently
  - Chlorine handling - 1 ton cylinders vs. a 90 ton rail car
    - Neighbor several kilometers away would consider the one ton cylinder inherently safer
    - Operators who have to connect and disconnect cylinders 90 times instead of a rail car once would consider the rail car inherently safer
  - Who is right?

44

Reducing risk or transferring risk?

- Reduce size of hazardous material storage tank at a plant
- Requires changing shipping mode from 150,000 Kg rail cars to 15,000 Kg trucks (smaller tank won't hold a rail car load)
- 10 X as many shipments, on road (more hazardous?) rather than on railroad
- Reduced site risk, possibly overall increased risk to society
- Supplier may have to maintain larger inventory at his plant

45

Holistic view of inherent safety

- Consider the full process life cycle and supply chain from raw materials to consumption and disposal
- *CONSIDER ALL HAZARDS!*
  - HAZARD IDENTIFICATION – You can't manage a hazard which you have not identified!
  - Informed decisions about conflicting goals
  - May be different choices for different situations
    - One floor houses eliminate risk of falling down stairs
    - So, why are many houses on a beach near the ocean built on stilts?
      - concern about a different hazard
- Think inherent safety at all levels of design and operation

46

Questions designers should ask when a hazard has been identified

Ask, in this order:

1. Can I eliminate this hazard?
2. If not, can I reduce the magnitude of the hazard?
3. Do the alternatives identified in questions 1 and 2 increase the magnitude of any other hazards, or create new hazards?
   (If so, consider all hazards in selecting the best alternative.)
4. At this point, what technical and management systems are required to manage the hazards which inevitably will remain? (layers of protection – passive, active, procedural)

47

National Academies Study: The Use and Storage of Methyl Isocyanate (MIC) at Bayer CropScience

- Report issued May 11, 2012
- Available for download from:

http://www.nap.edu/catalog.php?record_id=13385

Books and Other Resources

1. Kletz, T. A., and P. Amyotte, *Process Plants - A Handbook for Inherently Safer Design*, 2nd Edition, CRC Press, New York, 2010.
2. Center for Chemical Process Safety, *Inherently Safer Chemical Processes - A Life Cycle Approach*, 2nd Edition, American Institute of Chemical Engineers and John Wiley & Sons, New York, 2009.
3. *Guidelines for Engineering Design for Process Safety,* 2nd Edition. American Institute of Chemical Engineers and John Wiley & Sons, New York, 2012.
4. National Research Council. *The Use and Storage of Methyl Isocyanate (MIC) at Bayer CropScience* . Washington, DC: The National Academies Press, 2012.
5. Chapter on ISD in current edition of *Lees' Loss Prevention in the Process Industries.*
6. CCPS documents: Go to http://www.aiche.org/ccps/ and search the CCPS web site for "inherent"
7. Extensive literature in process safety journals. Literature on process intensification, green chemistry, nanotechnology is often relevant to inherently safer process design.

49

The Design of Safe, Efficient and Green Processes for API's Throughout Product Lifecycle

Dillon, J. L., Pfizer

Originally presented on: 6/6/2012 12:30:00 - 13:15:00

Manufacturing performance is a key component of Pfizer's effort in producing a "green product." The design of green processes for manufacturing Active Pharmaceutical Ingredients (API's) offers the greatest opportunity maximizing efficiency and reducing waste. In recent years a number of process tools, equipment and technologies have proven to be essential in augmenting our effort toward developing efficient, safe and green processes. Of particular importance at Pfizer is the use of Best Process Chemistry, BRITEST, Design of Experiments, High Throughput Reaction Screening, specialized facilities and the application of flow processes and bio-transformations, the latter of which are conducted in aqueous media. A case study demonstrating our approach to process design and optimization for maximizing API manufacturing performance throughout the lifecycle of a marketed product will be presented.

Use of Piloting to Reduce Scale-Up Risk

David A. Berg, Susan M. Hennessey, Kenneth E. Mersman, and Kathrine A. Threefoot

E.I. du Pont de Nemours and Company

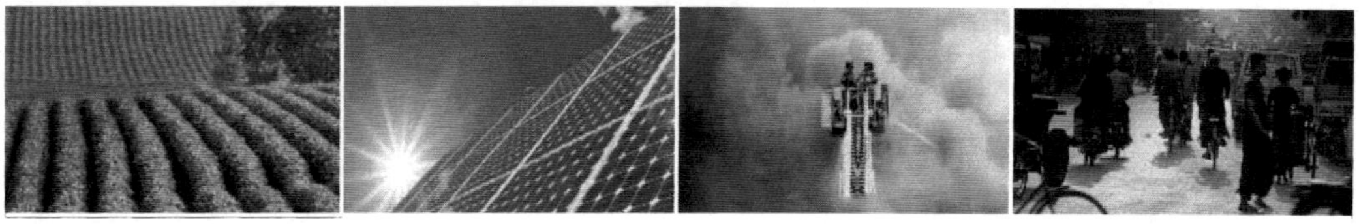

1

Presented at AIChE Process Development Symposium
June 6, 2012

Scale-Up Risks and Uncertainties

- **New Products:**
  - ✓ Customer Critical-to-Quality (CTQ) requirements
  - ✓ Process capability to meet CTQs
  - ✓ Market growth rate
- **Plant Operation:**
  - ✓ Attainment of nameplate capacity
  - ✓ Overall product yield
  - ✓ Operational challenges (fouling, plugging, corrosion, etc.)
  - ✓ Impact of recycle (build-up of impurities)
  - ✓ Process dynamics and control problems

2

You always build a pilot plant...

Some are at commercial scale

3

You always build a pilot plant...

Some are at commercial scale

Decisions about whether and how to pilot need to be based upon a thorough risk assessment

4

Risk Management Process

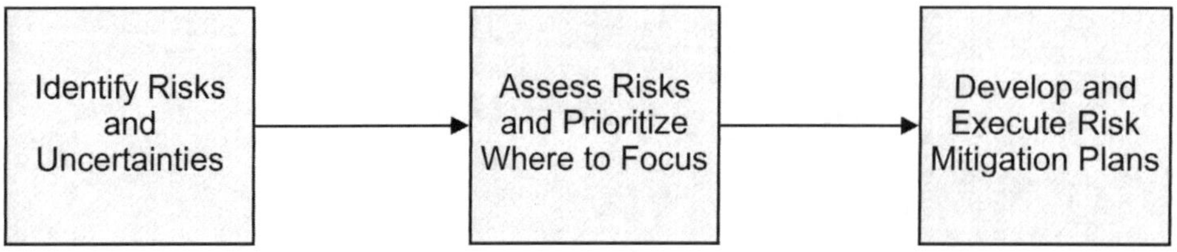

5

Risk Management Process

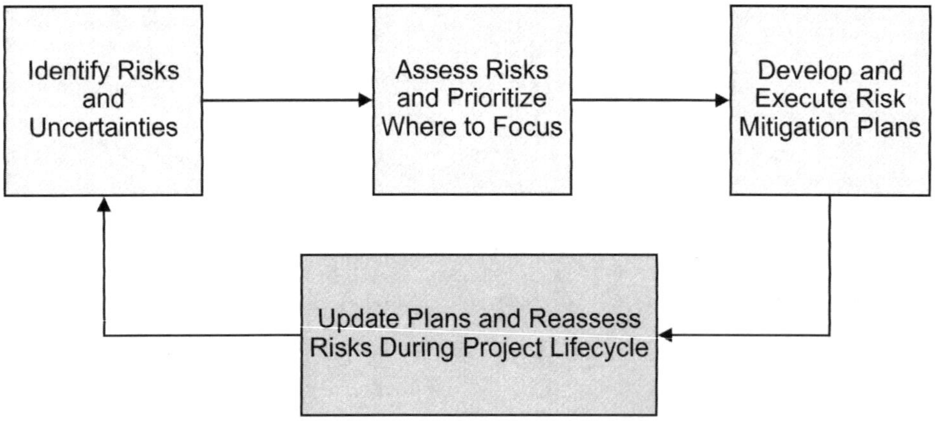

Risk Management Needs to be an Evergreen Process – not an Event!

6

Identification and Prioritization of Technology Risks

- **Engage a peer group external to the project team:**
  - ✓ Provide an unbiased perspective
  - ✓ Solicit input from subject matter experts
- **Follow a structured process:**
  - ✓ Review process technology – then discuss/capture risks
  - ✓ Use trained facilitators
  - ✓ Prioritize risks based on likelihood of event and severity of consequence (High, medium, and low)
- **Up-front preparation is key:**
  - ✓ Circulate technology package prior to review

7

Develop and Execute Risk Mitigation Plans

- **Focus initially on critical issues:**
  - ✓ High likelihood of occurrence
  - ✓ High severity of consequence
- **Manage as an evergreen process – not an event:**
  - ✓ Basis for deciding what work to do
  - ✓ Update plans regularly
  - ✓ Reassess risks at key decision gates

It's critical that Business Leaders participate in the Risk Management process

The Business needs to accept the degree of risk, and make risk-informed decisions

8

Piloting and Risk Management

- **Piloting is frequently a key element of most mitigation plans:**
  - ✓ Challenge is often justifying cost vs. benefit
  - ✓ IPA studies can be used to help justify need for piloting:
    - Reduce start-up time
    - Increase production in months 7-12 after mechanical completion
- **Key considerations:**
  - ✓ What process steps to pilot
  - ✓ Pilot plant scale
  - ✓ Where to pilot

9

Different Approaches to Piloting

Non-Integrated Piloting

- ✓ **Incorporates only 1-2 process steps** (e.g. chemical reactor)
- ✓ Typically smaller scale and operated for shorter periods of time
- ✓ Sometimes referred to as a Process Development Unit (PDU)
- ✓ Best used for process development and optimization

Integrated Pilot Plant

- ✓ **Incorporates all key steps and recycle loops**
- ✓ Typically larger scale and operated for longer periods of time
- ✓ Sometimes referred to as a Market Development Unit (MDU)
- ✓ Used for process demonstration and pre-commercial sampling

Vendor Equipment Tests

- ✓ Used for testing of specialized equipment (e.g. solids processing)
- ✓ Scale and scope of testing usually limited by raw material supply

10

Deciding What to Pilot

- **Review plans early and often:**
  - ✓ Evergreen approach to risk management
- **Base pilot plant design on vision for commercial plant:**
  - ✓ Scale-down vs. scale-up
- **Use a full integrated pilot plant for:**
  - ✓ Handling or processing of solids
  - ✓ Recycle streams
- **Design with the expectation of change:**
  - ✓ Deal with unanticipated problems
  - ✓ Accommodate new technology improvements
  - ✓ Build as small as possible to meet needs

11

Deciding the Pilot Plant Scale

Individual Unit Operations:

Many interdependent process variables - most are nonlinear

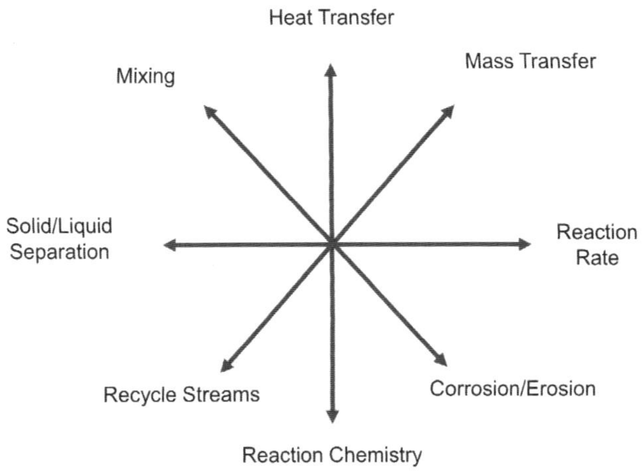

Which parameter do you use to scale-up?

Key is to decide what variables have the greatest impact to the process step

12

Deciding the Pilot Plant Scale (continued)

Average Scale-Up Ratios

Process	Lab to Pilot Plant	Pilot Plant to Commercial
All Gas	500-1000	200-1500
Gas Reactants Liquid/Solid Products	200-500	100-500
Liquid/Gas Reactants Liquid Products	100-500	100-500
Liquid Reactants, Solid Viscous Products	20-200	20-250
Solid Reactants, Solid Products	10-100	10-200

Bisio & Kabel, 1985

13

Deciding the Pilot Plant Scale (continued)

May need to do work at multiple scales based on:
- Scale-up considerations
- Business purpose for piloting

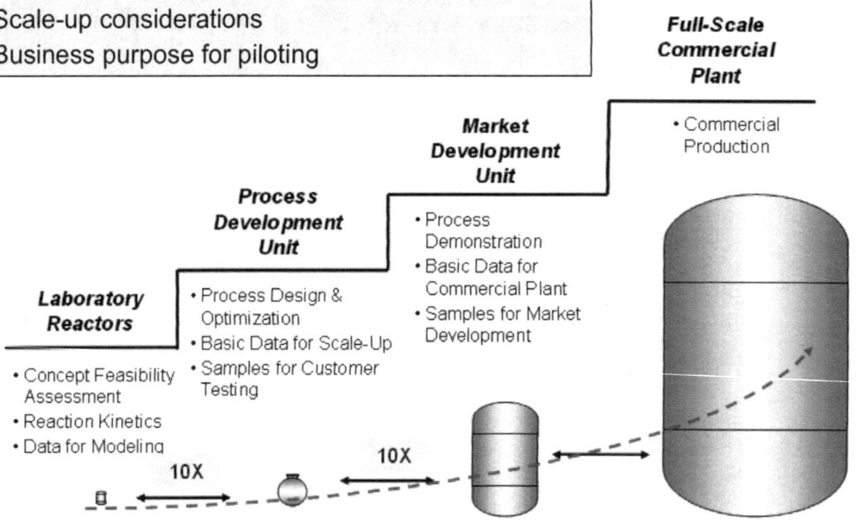

Approach to Commercialization for Cerenol® Polyols

14

Coordination and Planning of Parallel Activities

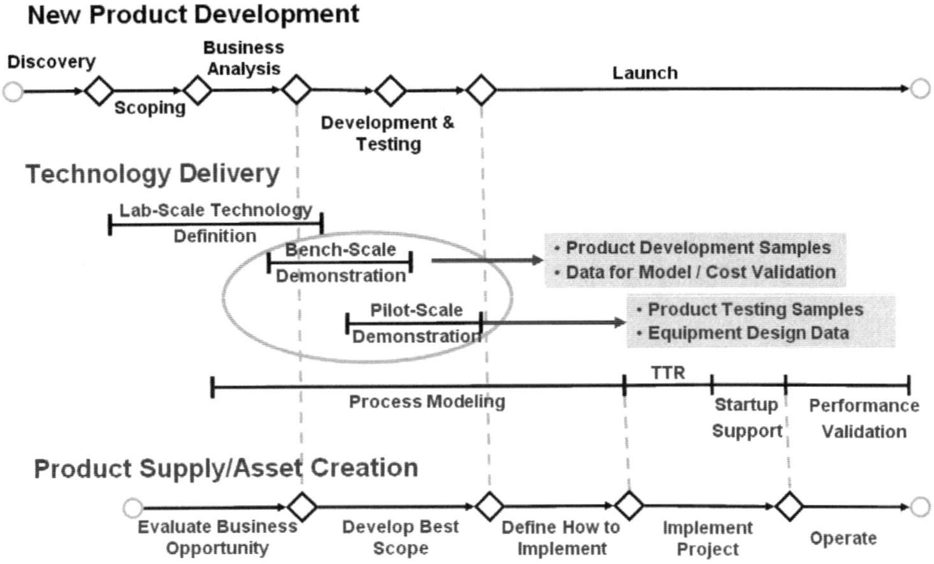

15

Use Process Modeling to Accelerate Piloting

Example - Development of Versipol® LAO technology

Develop Catalyst and Process Fundamentals
- Catalyst design
- Screening new catalyst
- Fundamental process model
- Laboratory reactor

grams/hr

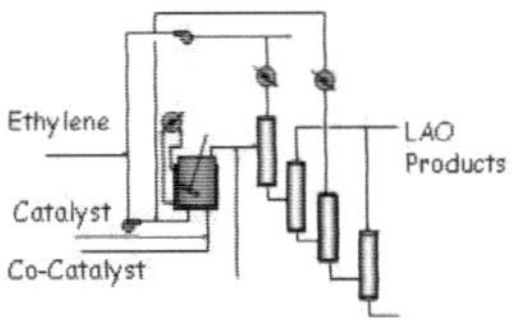

Process and Product Development
- Pilot plant operation
- Determine/validate process
- Operating window
- Select preferred catalyst
- Produce product samples

kg/hr

Scale-Up to Commercial Operation
- Use validated process model for design
- Detailed design
- Operating procedures
- Continue pilot plant tests to support start-up
- Start-up and commissioning

tonne/hr

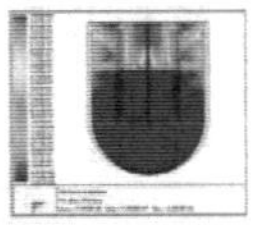

Process Modeling

16

Other Best Practices to Consider

- **Use the same feedstock as for the commercial plant**
- **Make all intermediates and products to the planned commercial specifications:**
  - ✓ Have customers evaluate product from the pilot plant
- **Keep the pilot plant available during commercial start-up:**
  - ✓ May be necessary to help resolve problems
- **Plan up-front for technology transfer:**
  - ✓ Identify basic data needs for commercial plant
  - ✓ Document what you learn
  - ✓ Use piloting team to support commercial start-up

17

Summary – Key Messages

- **Make risk management a part of the way you work:**
  - ✓ An evergreen process – not an event
  - ✓ Engage experts outside of project team
- **Start with a vision of the commercial process:**
  - ✓ Pilot plant design should be scaled down from this
- **Spend lots of time planning:**
  - ✓ Understand the business need
  - ✓ Decide which parts of the process to pilot
  - ✓ Determine what are critical scale-up parameters
- **Plan up-front for technology transfer:**
  - ✓ Capture all data and observations
  - ✓ Use piloting team to support commercial start-up

18

Use of Piloting to Reduce Scale-up Risk

The miracles of science™

Design and Scale Up of Fully Continuous Processes including Reaction, Workup, and Isolation

Johnson, M., Eli Lilly and Company

Originally presented on: 6/6/2012 15:00:00 - 15:30:00

Several fully continuous processes have been run in Chemical Product Research and Development at Eli Lilly to enable scale up of chemistries that are not suitable for conventional batch equipment. Superheated thermal cyclization, high pressure hydroformylation (70 bar), high pressure asymmetric hydrogenation (70 bar), which were not feasible to scale up in existing in-house batch equipment, have been conducted safely in continuous reactors. Multi-stage counter-current extractive workup enabled higher product yield and better impurity rejection compared to batch. Continuous crystallization enabled higher ee upgrade compared to batch, seeding only once at the beginning of the campaign, and the quality assurance of steady state operation. Continuous distillation enabled difficult solvent exchanges with less solvent waste and less residence time in the evaporator compared to batch, and efficient separation of compounds with small relative volatility. Typical research scale throughputs are 1 to 10 g/h and pilot scale throughputs are 5 to 15 kg/day. Most continuous processes that were run at research and pilot scale in the development laboratories have solids in flow. Custom methods and devices for slurry mass transport are described. No mechanical pumps are used for slurry flow. Instead, automated sequenced block valves and pressure swing fluid transfer chambers are used for pumping and flow metering of slurries.

EASTMAN

The Search for the Piloting and Scale-up Vision of the Future

Brad Duckworth, Technology Group Leader, Eastman Chemical Company
AIChE Process Development Symposium
King of Prussia, PA
June 5-7, 2012

Piloting Defined

- **Piloting** (and/or **Scale-up**): the process of scaling up a process or product concept from the idea to the commercial scale.

- **Piloting Purpose**: generate information useful for design of larger facilities

EASTMAN

Many Phases of Piloting Process

- **Phases commonly referenced with example scale factors (scale relative to commercial facility)**

Lab Scale 1:10,000-100,000

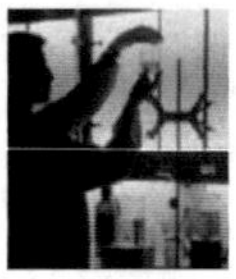

Bench Scale 1:1,000-10,000

Pilot Scale 1:100-1,000

Semi-Works Scale 1:10-100

Commercial Scale 1:1

3

EASTMAN

Why is Piloting Important?

4

Why is Piloting Important

- Process Improvement / Development
  Yes, but…NOT ENOUGH

- Companies MUST have NEW GROWTH!
  - New Products
  - New Processes
  - New Services

- How?? Growth can come in 2 major ways
  - INORGANIC: acquisition, license or purchased technology/process
  - ORGANIC: develop NEW in house

EASTMAN

Organic Growth Formula

Organic Growth = f (x)

- Where X can be many things such as…
  - Idea generation
  - Research & Development competency
  - R&D budget
  - Sales & Marketing competency
  - Business Strategy
  - Innovation Process

- **MAJOR X = Piloting & Scale-up Competency**

6

EASTMAN

Strong Piloting Competency Pays Dividends for Growth

- Faster Growth
  - Reduced rework means getting to market faster
- More Growth
  - Higher success rate with taking ideas/concepts to commercial scale means more market share
- Lower $$/Growth
  - More $ and resources to work on more opportunities means more markets touched

- Summary: Piloting and Scale-up excellence can significantly improve Growth & Market Penetration!!

7

EASTMAN

EASTMAN

Product Life Cycle

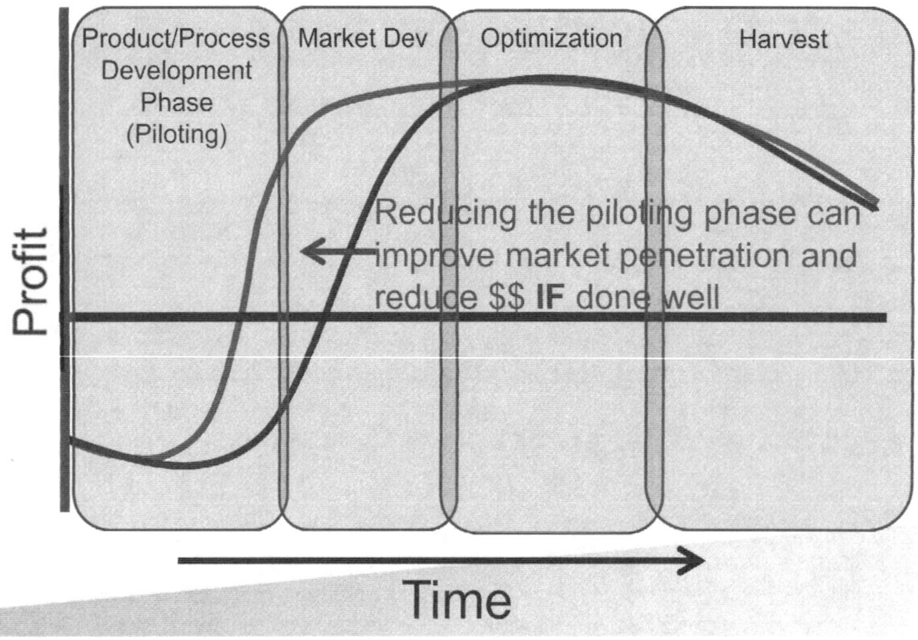

8

Piloting Vision Team Charged

Piloting Vision Team Background

- Eastman Management saw need to be the best at Piloting

- Team charged: "Develop a VISION for how Piloting should be done at Eastman to maximize growth"

Approach Utilized

6 Sigma

11

Business Design for 6 Sigma Roadmap

EASTMAN

Define

- Initiate the Project
- Identify Target Process Stakeholder
- Define Business Needs
- Gather Detailed Stakeholder Requirements

Conceptualize

- Create Initial Process Concepts
- Select Best Concept Hybrid
- Select Key Process Output Variables

Design

- Identify Detailed Process Specifications
- Create Detailed Process Maps
- Understand Process Inputs
- Model Critical Processes
- Develop and/or Evaluate Measurement Systems

Optimize

- Minimize Process Complexity, Maximize Entity Velocity
- Optimize Critical Inputs: Process Simulation

Control

- Finalize the Control System
- Verify Long Term Capability

Business Design for 6 Sigma Roadmap

EASTMAN

Homework	Define	Conceptualize	Design	Optimize	Control
	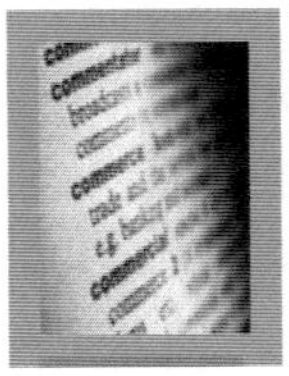				
❑ Internal Capabilities ❑ External Benchmarking	❑ Initiate the Project ❑ Identify Target Process Stakeholder ❑ Define Business Needs ❑ Gather Detailed Stakeholder Requirements	❑ Create Initial Process Concepts ❑ Select Best Concept Hybrid ❑ Select Key Process Output Variables	❑ Identify Detailed Process Specifications ❑ Create Detailed Process Maps ❑ Understand Process Inputs ❑ Model Critical Processes ❑ Develop and/or Evaluate Measurement Systems	❑ Minimize Process Complexity, Maximize Entity Velocity ❑ Optimize Critical Inputs: Process Simulation	❑ Finalize the Control System ❑ Verify Long Term Capability

Homework-Internal Capabilities

14

Homework-Internal Capabilities

EASTMAN

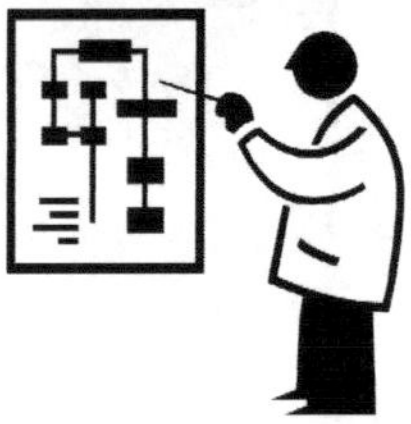

- Equipment
  - What do we have?
  - Where is it?
  - How is it being used?
- Processes
  - How is piloting done
  - Best practices
- Personnel
  - Selection
  - Core Competency
  - Retention

15

Snapshot of Capability Findings -Equipment

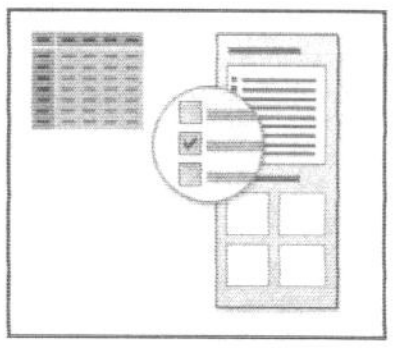

- Developed Piloting Equipment Database
  - Catalogued >600 major equipment items or bench units in 17 different buildings (Kingsport and Longview).
  - Items in Database Characterized and Searchable by:
    - Scale, Material of Construction, Flexibility, etc.
  - Data Analyzed for trends

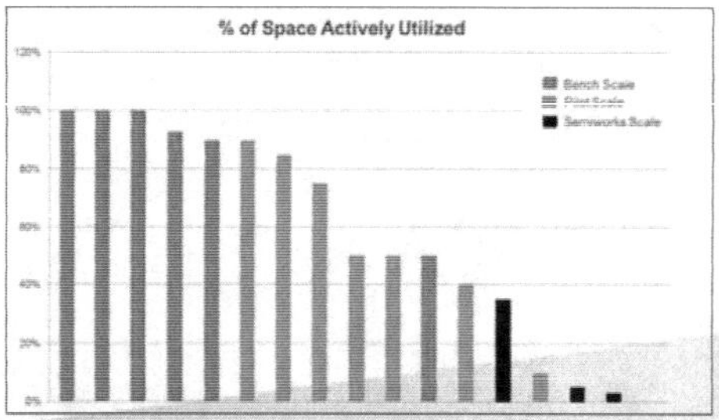

Individual Piloting Units

16

EASTMAN

Sample of Internal Capability Findings

- Equipment
  - Knowledge of Equipment Capability siloed
  - Safety restrictions on use of equipment exist
- Processes
  - Scale-up methodology not consistent
  - "Best practices" exist
- Personnel
  - Recruiting and placement of personnel solid
  - Management of Piloting core competency needs focus

17

EASTMAN

Homework-Benchmarking

18

Benchmarking-2 Parts

- Part 1: Broad industry benchmarking
  - Gained non-confidential piloting information from various companies
    - Internet
    - Trade press
    - Employee benchmarking interviews
- Part 2: Expert/Consultant benchmarking
  - Hired/Interviewed consultants who are considered piloting experts by industry
    - Interviews
    - Classes

19 EASTMAN

Sample of Benchmarking Findings

- Process
  - Detailed scale-up process paramount
  - Early focus on kinetics, modeling utilized more than EMN
  - Data quality is critical early in scale-up, EMN struggles here
- Assets
  - Open bay and modular utilized more than EMN
- Personnel
  - Highly qualified staffing targeted for piloting
- External
  - Must be part of piloting program but not 1st consideration

20 EASTMAN

Homework DONE!

EASTMAN

 Homework	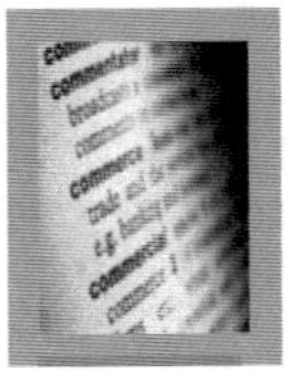 Define	 Conceptualize	 Design	 Optimize	 Control
❑ Internal Capabilities ❑ External Benchmarking	❑ Initiate the Project ❑ Identify Target Process Stakeholder ❑ Define Business Needs ❑ Gather Detailed Stakeholder Requirements	❑ Create Initial Process Concepts ❑ Select Best Concept Hybrid ❑ Select Key Process Output Variables	❑ Identify Detailed Process Specifications ❑ Create Detailed Process Maps ❑ Understand Process Inputs ❑ Model Critical Processes ❑ Develop and/or Evaluate Measurement Systems	❑ Minimize Process Complexity, Maximize Entity Velocity ❑ Optimize Critical Inputs: Process Simulation	❑ Finalize the Control System ❑ Verify Long Term Capability

Define Phase

22

Stakeholder Requirements

- Stakeholders
  - R&D, Manufacturing, Engineering, Business, Marketing, Finance
- Interviews
  - Voice of the Customer questionnaire
  - Over 2000 voices collected
- Key Requirements
  - Workshop
  - KJ

23 EASTMAN

Piloting Vision "Requirements" KJ Diagram

EASTMAN

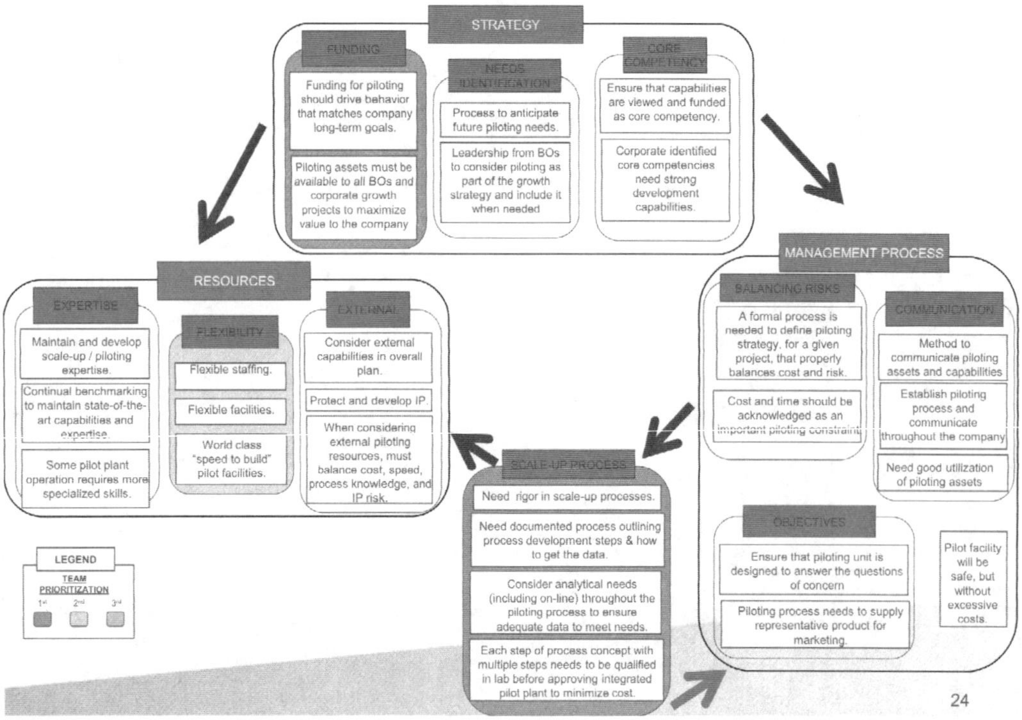

24

Validated: Top 10 Key Requirements

EASTMAN

- Requirements development based on Stakeholder VOCs
- Requirements validated and ranked through survey based on 1-5 scale with 1=not important and 5=Most important

25

Homework + Define DONE!

EASTMAN

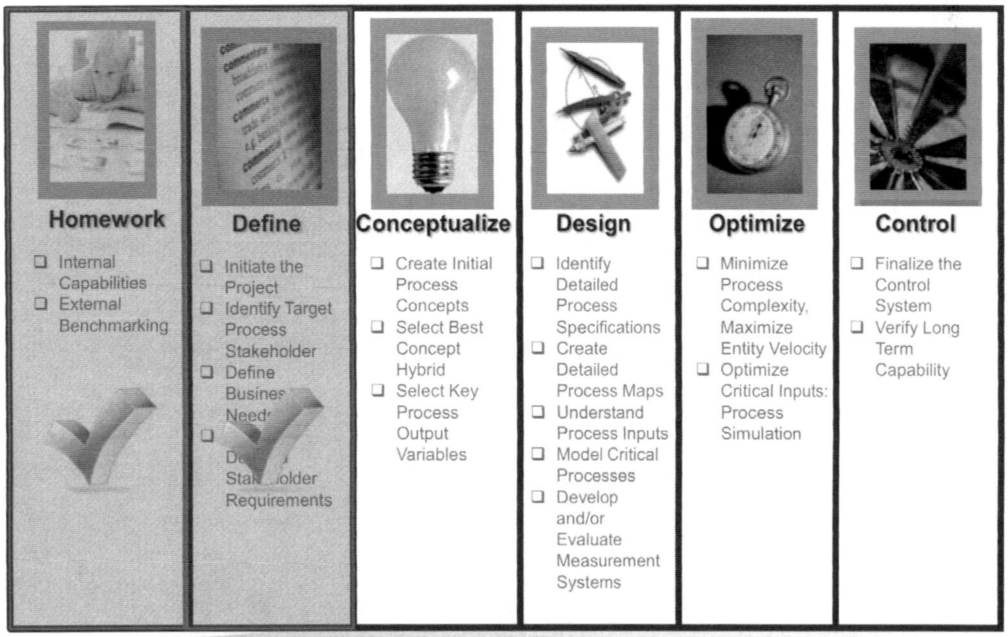

Developed Solutions to Meet Requirements

Conceptualize

- Brainstormed Ideas
- Developed Concepts
- Pugh Concept Selection / Hybridization

27

EASTMAN

EASTMAN

Key Solution Concepts for Piloting Vision

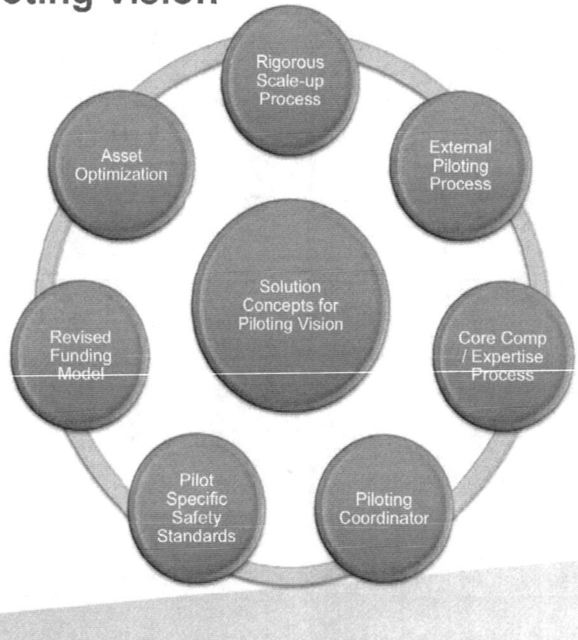

28

EASTMAN

Key Solution Concepts for Piloting Vision

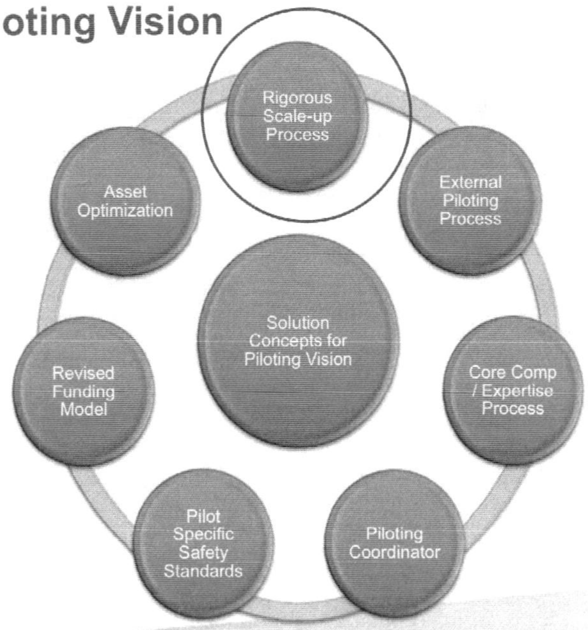

29

Scale-up Process: What is it?

- The Scale-up Process is a <u>technology innovation workflow</u> including a framework and tools to provide the deliverables for Eastman's Stage Gate process
- Case for Change
  - Each Eastman organization does scale-up differently
  - Eastman scale-up process has gaps
- Key Benefit: Scale-up is not cheap, if we are not optimized we are losing money and sacrificing growth

30

EASTMAN

Scale-up Framework & Tools

Alignment of Stage Gate , Six Sigma for New Product and Process Innovation

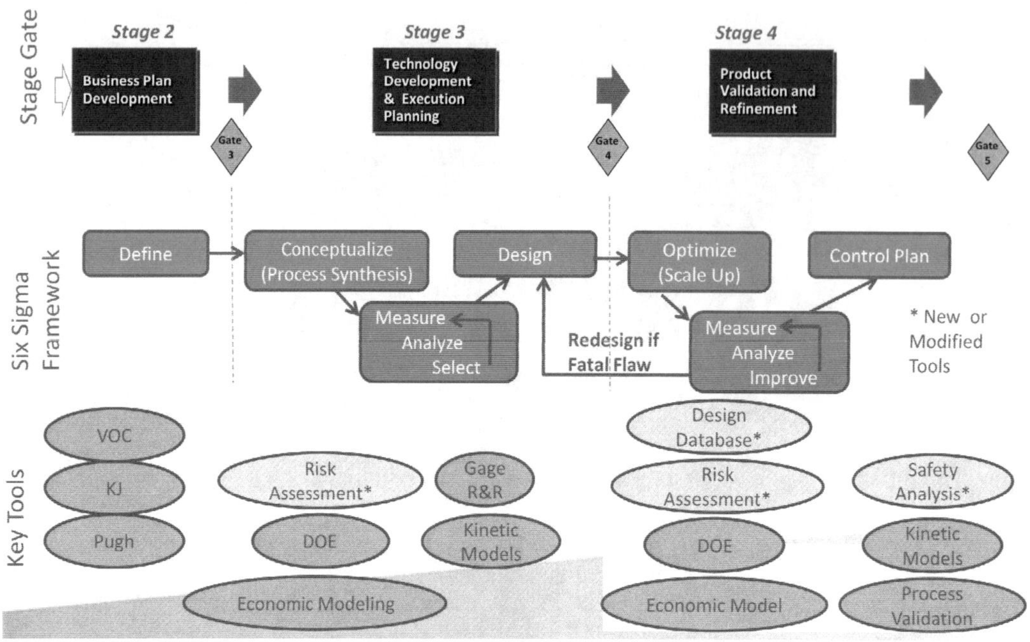

Key Solution Concepts for Piloting Vision

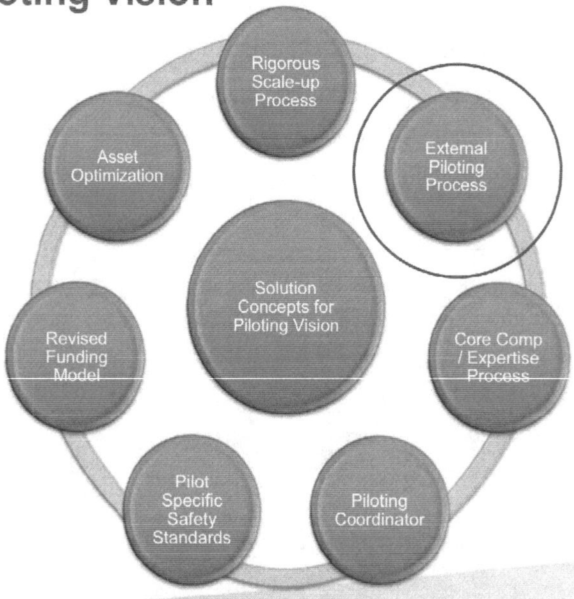

32

External Piloting Process: What is it?

- A process for managing when and how to utilize external piloting capability
- Case for Change
  - Benchmarking suggests that others include as key piece of development program
  - Can bring value to Eastman if done right, can cost a lot if done wrong
- Key Benefit: Collaboration drives Innovation

33 EASTMAN

External Piloting Process

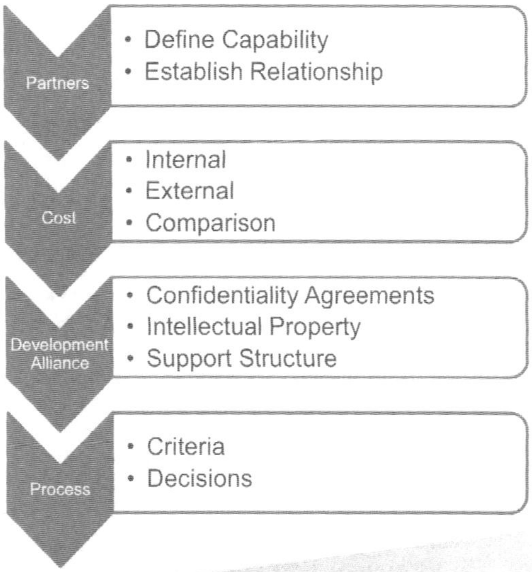

34 EASTMAN

Key Solution Concepts for Piloting Vision

EASTMAN

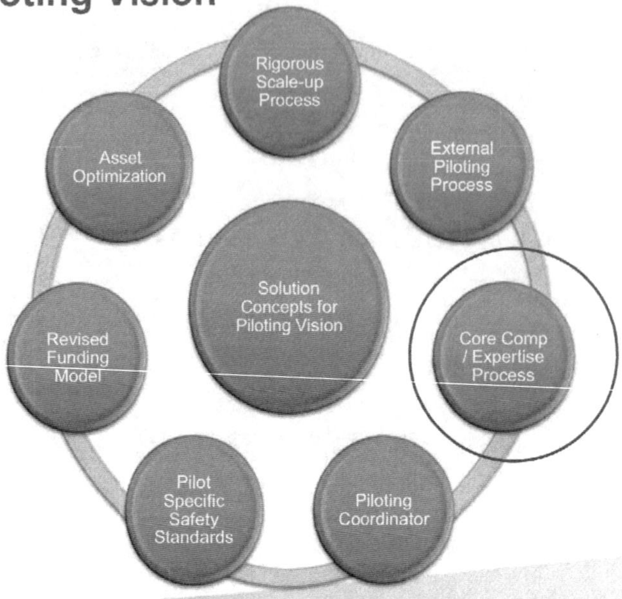

35

Core Comp Process: What is it?

- A process for developing and managing piloting and scale-up core competency and expertise
- Case for Change
  - Benchmarking suggests industry is losing expertise and Eastman little different
  - Unique technology differentiates-you must have expertise to handle complexities
- Key Benefit: Competitive Advantage

36

EASTMAN

Piloting Core Competency Process

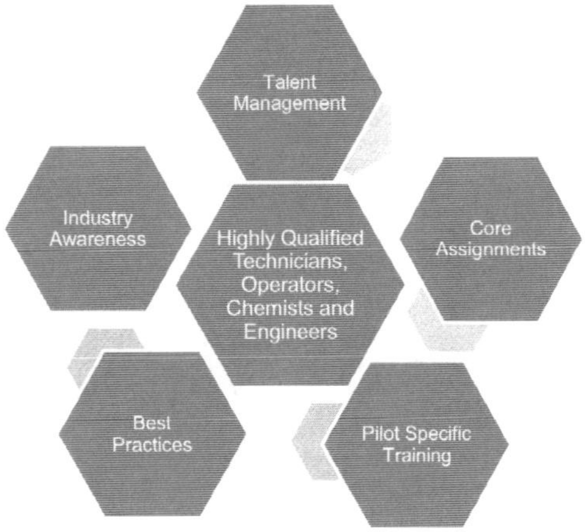

EASTMAN

37

Key Solution Concepts for Piloting Vision

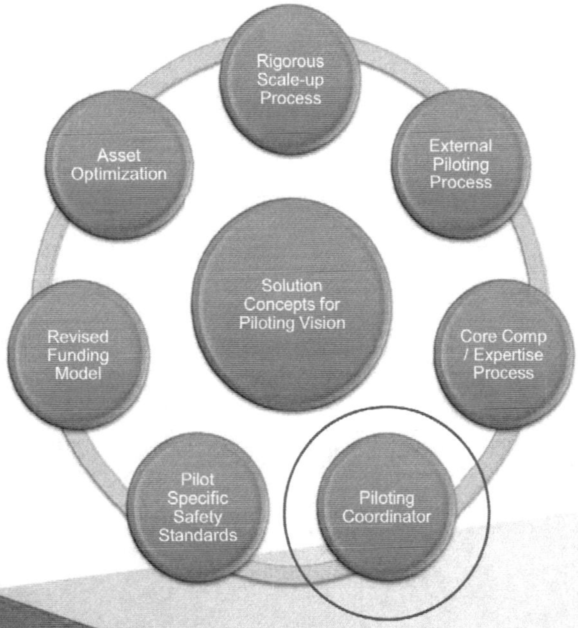

EASTMAN

38

Piloting Coordinator: What is it?

- A single individual assigned to be the focal point for all Piloting at Eastman
- Case for Change
  - Need focal point for Business Leadership
  - Needed to run new processes
  - Internal coordinator prototype a success
- Key Benefit: Champion

39

EASTMAN

Piloting Coordinator: What does it look like?

Piloting Coordinator

Chief Technology Officer
+ Innovation Council

⇧

Pilot Plant Networking Team

Membership
Pilot Plant Group Leaders
Piloting & Technical Experts

40

EASTMAN

EASTMAN

Key Solution Concepts for Piloting Vision

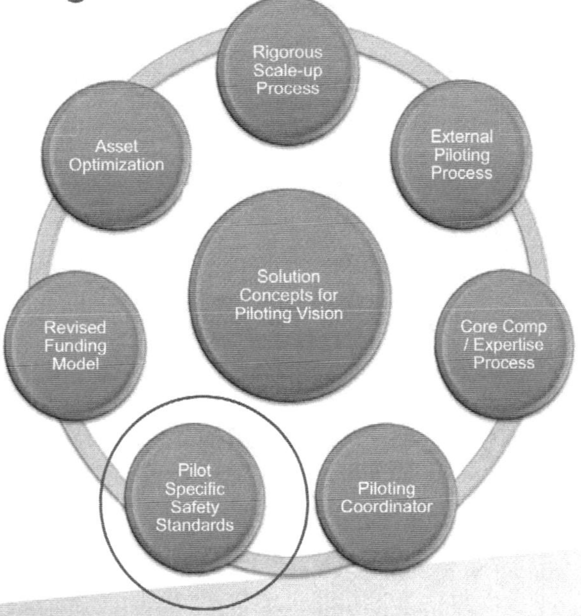

41

Pilot Specific Safety Standards: What is it?

- Safety Standards customized specifically for bench and pilot scale equipment and work
- Case for Change
  - Internal engineering standards are written with plant operations in mind, often not specific to smaller flowrates of bench and pilot scale equipment
- Key Benefit: *Relevant* designs quicker AND inherently safer

42

EASTMAN

Key Solution Concepts for Piloting Vision

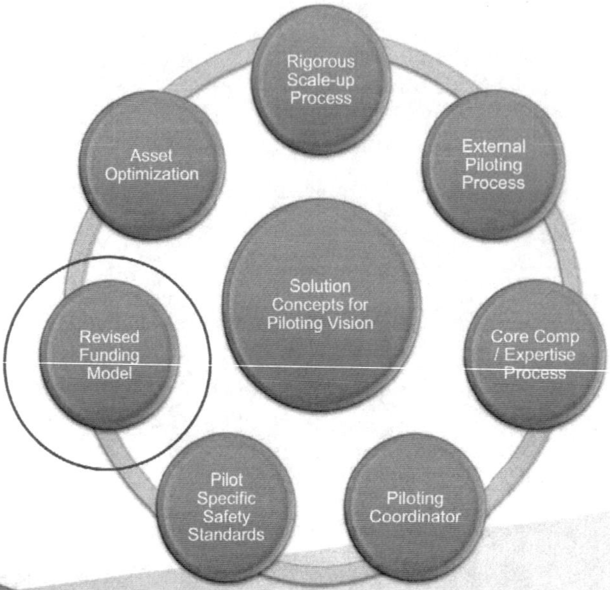

43

EASTMAN

Revised Funding Model: What is it?

- A change in how pilot plants <u>costs are distributed</u> and how piloting <u>funding decisions</u> are made
- Case for Change
  - Need to Change the conversation with Business Leaders
    - From "Will you fund piloting next year…" to
    - "How can we use piloting to support your strategy"
  - Difficult to maintain adequate asset and staffing models
  - Drive behaviors aligning with longer term growth strategy
- Key Benefit: Funding and related decisions more aligned with longer term growth strategy of company

44

EASTMAN

Funding Model: 3 Major Revisions

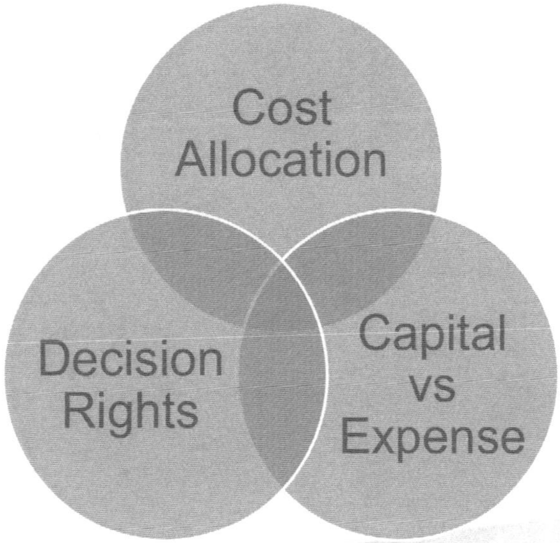

45

Example: Decision Rights

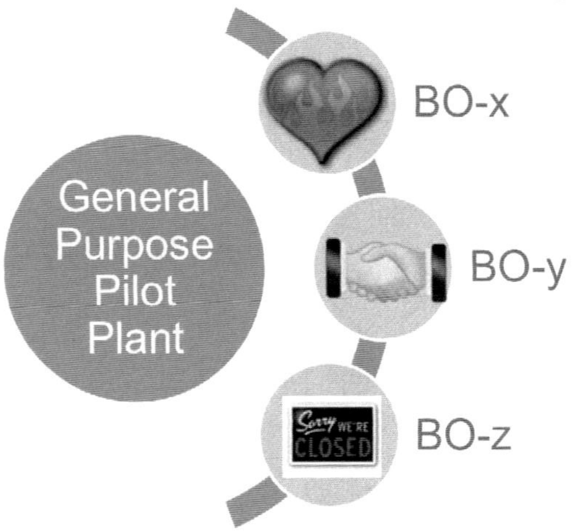

Overall Funding and Strategy Decisions

- Longer Term consistency in resources & expertise
- Strategic

Decision Rights MUST be elevated- Innovation Council

- Short Term
- Tactical

46

Key Solution Concepts for Piloting Vision

EASTMAN

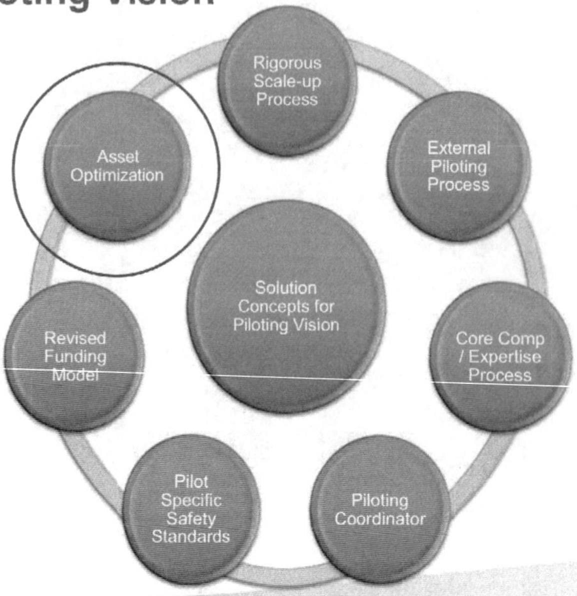

47

Asset Optimization: What is it?

- Eastman moves from the current method of piloting (field constructed assets that we try to modify and reuse as much as possible) to open bay modular piloting facilities designed specifically for need
- Case for Change
  - Current "field construction and reuse" method too costly and slow
  - Benchmarking suggests industry moving to this format
- Key Benefit: Speed

48

EASTMAN

A Modular Piloting Facility (Bench to Pilot)

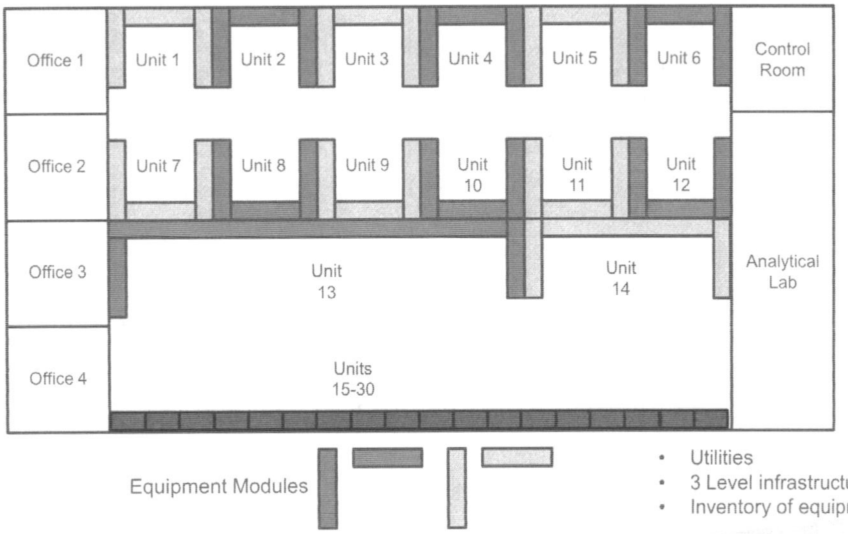

- Utilities
- 3 Level infrastructure
- Inventory of equipment

EASTMAN

49

Examples of Modular Piloting

Photo credit to:
-chemicalprocessing.com
-pdcmachines.com
-zeton.com

50

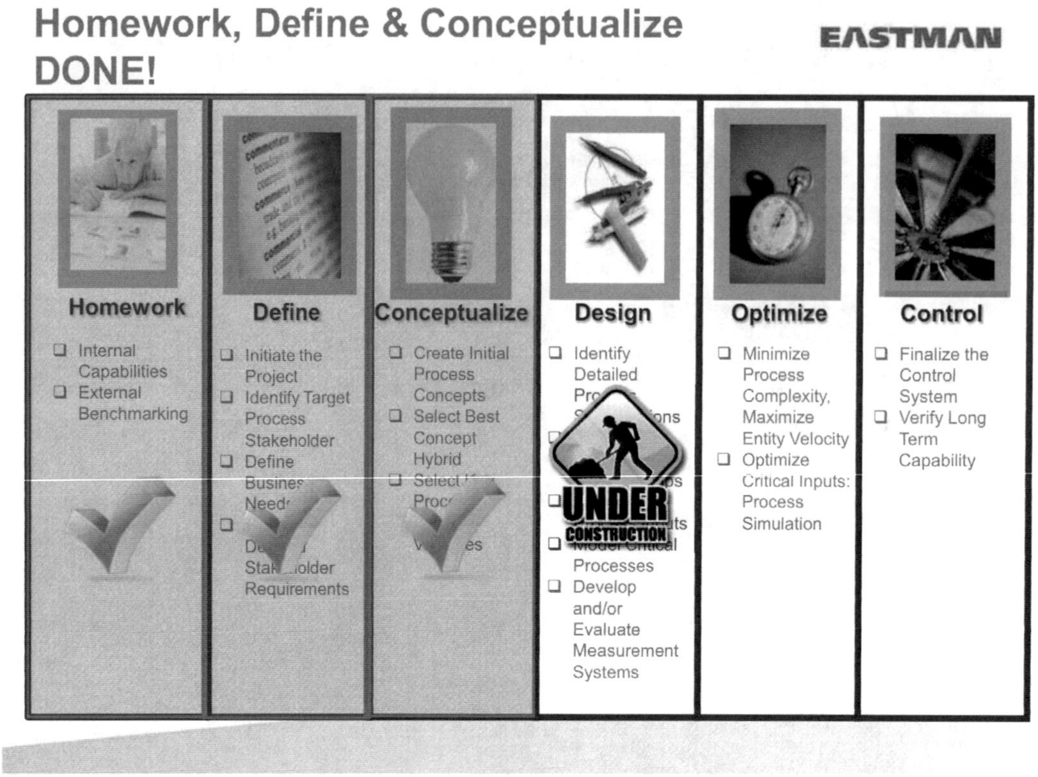
Homework, Define & Conceptualize DONE!
EASTMAN
Homework
Internal Capabilities
External Benchmarking
Define
Initiate the Project
Identify Target Process Stakeholder
Define
Stakholder Requirements
Conceptualize
Create Initial Process Concepts
Select Best Concept Hybrid
Design
Identify Detailed
UNDER CONSTRUCTION
Processes
Develop and/or Evaluate Measurement Systems
Optimize
Minimize Process Complexity, Maximize Entity Velocity
Optimize Critical Inputs: Process Simulation
Control
Finalize the Control System
Verify Long Term Capability

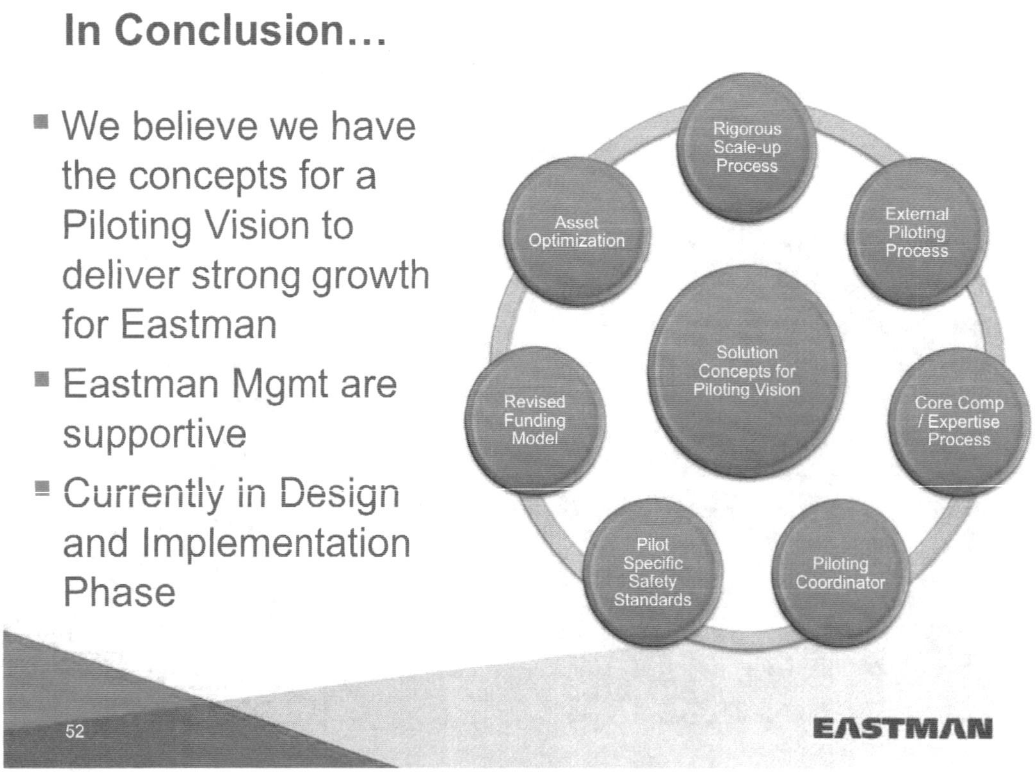
In Conclusion…
We believe we have the concepts for a Piloting Vision to deliver strong growth for Eastman
Eastman Mgmt are supportive
Currently in Design and Implementation Phase
Rigorous Scale-up Process
External Piloting Process
Core Comp / Expertise Process
Piloting Coordinator
Pilot Specific Safety Standards
Revised Funding Model
Asset Optimization
Solution Concepts for Piloting Vision
52
EASTMAN

Thanks for your time!

Questions??

Using Physical Properties to Develop Better Chemical Products and Processes

Dr. Kevin G. Joback
Molecular Knowledge Systems, Inc.
kevin@molecularknowledge.com

Goal

"People don't need drills. People need holes" – Basic Marketing

"People don't need chemicals. People need the <u>properties</u> those chemicals possess" – Kevin G. Joback

To have you <u>appreciate the importance of physical properties</u> in all your decisions regarding process and product development.

2000-11-15-01

Outline

Why Physical Properties ?

1. **Separation Problem:** fake gold bars
2. **Property Constraints:** categories, lifecycle
3. **Candidate Generation:** states, combine
4. **Evaluate:** computations, experiments

2000-11-15-01

Molecular Knowledge Systems, Inc. *Page 3*

Separation Problem

Real ?
Fake ?

Typical Gold Bar

Weight = 400 troy oz

Weight = 12.441 kg

Size = 25 x 6 x 4.3 cm

Density = 19.30 g/cc

Purity = 99.5 wt%

Value = $639,248

2000-11-15-01

Molecular Knowledge Systems, Inc. *Page 4*

Density

Density = 19.30 g/cc – denser than most metals

$$\rho_{mix} = \rho_A \times wt_A + \rho_B \times wt_B$$

Element	Density	$ / kg
Bismuth	9.78	10
Tungsten	19.25	24
Gold	19.30	51,382
Iridium	22.56	13,150
Osmium*	22.59	19,600

* health and manufacturing issues

Tungsten	98.49 %
Iridium	1.51 %
Density	19.30 g/cc
Cost	222 $/kg

2000-11-15-01

Heat Capacity

Element	J/g K
Bismuth	0.122
Gold	0.129
Iridium	0.131
Tungsten	0.134

100°C

\+

20°C

$$m_G C_{p,G}(100 - T_f) = m_W C_{p,W}(T_f - 20)$$
$$T_f = 42.19$$

$$\rho_{mix} = \sum \rho_i wt_i$$
$$C_{p,mix} = \sum C_{p,i} wt_i$$

Iridium	70.4 %
Bismuth	24.1 %
Tungsten	5.5 %
Heat Cap	0.129 J/g K
Density	19.30 g/cc
Cost	9,261 $/kg

2000-11-15-01

Look and Feel

Element	Mohs	Color
Bismuth	2.25	gray
Gold	2.50	yellow
Iridium	6.50	silver
Tungsten	7.50	silver

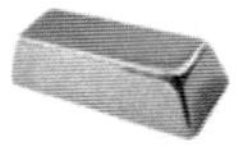

vs.

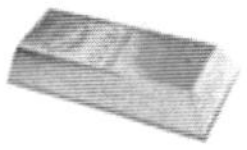

Gold Plating

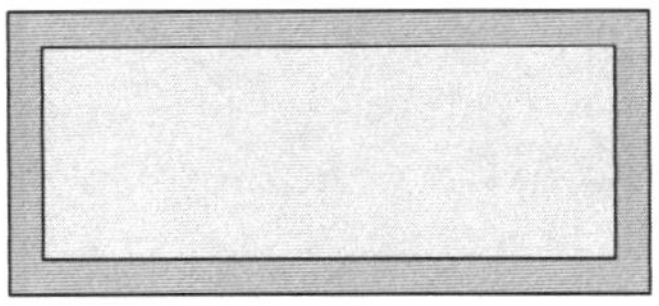

0.2 cm thick

Iridium	58.6 %
Bismuth	20.1 %
Tungsten	4.6 %
Gold	16.7 %
Cost	16,290 $/kg

2000-11-15-01

Acoustic Impedance

Ultrasound – Boundary Echoes

Sound Reflection

$$R = \left[\frac{\rho_1 c_1 - \rho_2 c_2}{\rho_1 c_1 + \rho_2 c_2} \right]^2$$

Element	m / sec
Gold	3240
Tungsten	5180
Iridium	4825

$c = f(x)?$

Remove Boundaries

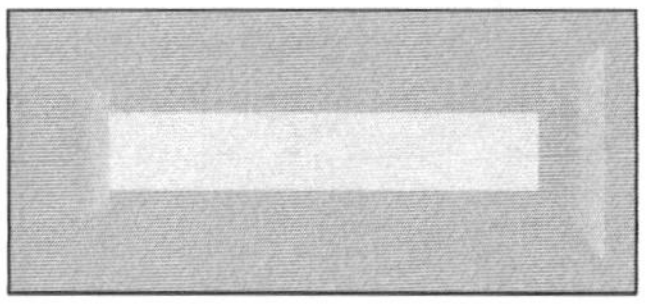

Cost ?

2000-11-15-01

Example Review

Three Step Approach

1. **Developed Property Constraints:**
   - performance, look, feel, safety, manufacturability
2. **Generated Candidates:**
   - chemicals, mixtures, structure
3. **Evaluated Candidates:**
   - computations, experiments

Past Projects

- Artificial Blood
- Refrigeration Lubes
- Degreasing Solvents
- Non-MMH Rocket Fuels
- Aircraft Deicing Fluids
- Heat Transfer Fluids
- MeCl2 Replacement
- Sonar Fill Fluids
- Phase Change Materials
- CO2 Absorption Solvents
- Hydraulic Energy Storage
- Windshield Washer Fluid
- Shatterable Coatings
- Soil Consolidants

Common Thread: Physical Properties

Approach: Step 1

Three Step Approach

1. **Developed Property Constraints:**
   - Sources, Properties, Values
2. **Generated Candidates:**
   - chemicals, mixtures, structure
3. **Evaluated Candidates:**
   - computations, experiments

Fake Gold Bar

Property Constraints

1. ρ_{fake} = 19.30 g/cc
2. $C_{p,fake}$ = 0.129 J/g K
3. R = small as possible
4. Color = golden
5. Low toxicity

Physical Properties

- ❑ Melting Point
- ❑ Chemical Reactivity
- ❑ Densities
- ❑ Vapor Pressure
- ❑ Surface Tension
- ❑ Viscosity
- ❑ Thermal Conductivity
- ❑ Heat Capacities
- ❑ Biological O2 Demand
- ❑ Henry's Law Constant
- ❑ Chemical Stability
- ❑ Heat of Vaporization
- ❑ Odor Threshold
- ❑ Flammability Limits
- ❑ Water Solubility
- ❑ Toxicity

plus many more ...

2000-11-15-01

Property Categories

Organize into Categories

- ❑ Interfacial
- ❑ Reactivity
- ❑ Thermal
- ❑ Sensory
- ❑ Optical
- ❑ Health
- ❑ Safety
- ❑ Phase Equilibria
- ❑ Environmental
- ❑ Rheological
- ❑ Mechanical
- ❑ Diffusional
- ❑ Volumetric
- ❑ Electrical

2000-11-15-01

Sources of Constraints

Lifecycle Analysis

- ❑ **Raw Materials:** sources, hazards
- ❑ **Manufacture:** process limits
- ❑ **Shipping:** extreme conditions
- ❑ **Purchase:** look, feel
- ❑ **Use:** performance
- ❑ **Abuse:** toxicity, flammability
- ❑ **Disposal / Recycle:** degradation, separation

2000-11-15-01

Molecular Knowledge Systems, Inc. *Page 15*

Sources of Constraints: Shipping

New type of window screens using a new, ultrathin material. Shipped to Arizona for testing.

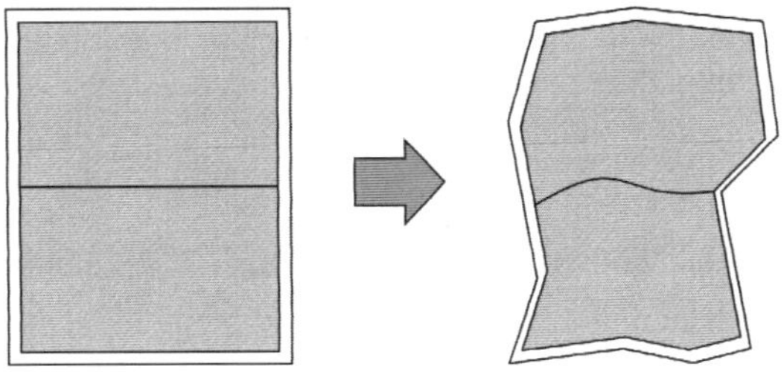

Rule of Thumb

The inside of a car or truck can reach 180F.

2000-11-15-01

Molecular Knowledge Systems, Inc. *Page 16*

Property Constraints Checklist

Category – Property – Life Stage

Volumetric Behavior

✓ **Coefficient of thermal expansion**
- Shipping: contents expand on temperature rise
- Use: mismatch of adhered materials may lead to cracks

✓ **Speed of sound**
- Use: must reduce acoustic reflection to near zero

✓ **Density**
- Use: density must match gold's 19.30 g/cc

❑ Volume of mixing

❑ Compressibility

❑ Volume

2000-11-15-01

Molecular Knowledge Systems, Inc. Page 17

Windshield Washer Fluid

❑ Raw Materials
❑ Manufacture
❑ Shipping
❑ Purchase
❑ Use
❑ Abuse
❑ Disposal

❑ Phase Equilibria
❑ Environmental
❑ Rheological
❑ Mechanical
❑ Diffusional
❑ Volumetric
❑ Interfacial
❑ Reactivity
❑ Electrical
❑ Sensory
❑ Thermal
❑ Optical
❑ Safety

2000-11-15-01

Molecular Knowledge Systems, Inc. Page 18

Property Constraints Checklist

Category – Property – Life Stage

Phase Equilibrium Behavior

✓ **Solid-Liquid Equilibria**
- Shipping: must not solidify above -20°C
- Use: must dissolve splattered bugs
- Abuse: must not dissolve automotive materials

✓ **Vapor-Liquid Equilibria**
- Shipping: boiling point greater than "back of truck temp"
- Use: must evaporate from windshield

❑ Liquid-Liquid Equilibria

❑ Henry's Law Constant

2000-11-15-01

Molecular Knowledge Systems, Inc. Page 19

Property Constraints Checklist

Category – Property – Life Stage

Volumetric Behavior

✓ **Volume**
- Purchase: concentrated packages save space
- Shipping: space costs money
- Examples: High Efficiency laundry detergents

❑ Coefficient of thermal expansion

❑ Volume of mixing

❑ Compressibility

❑ Speed of sound

❑ Density

2000-11-15-01

Molecular Knowledge Systems, Inc. Page 20

CO2 Absorption Solvent

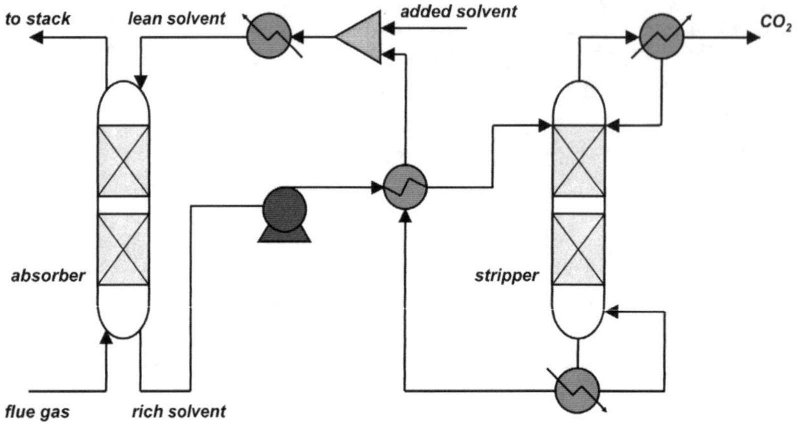

2000-11-15-01

Molecular Knowledge Systems, Inc.

Page 21

CO2 Absorption Solvent

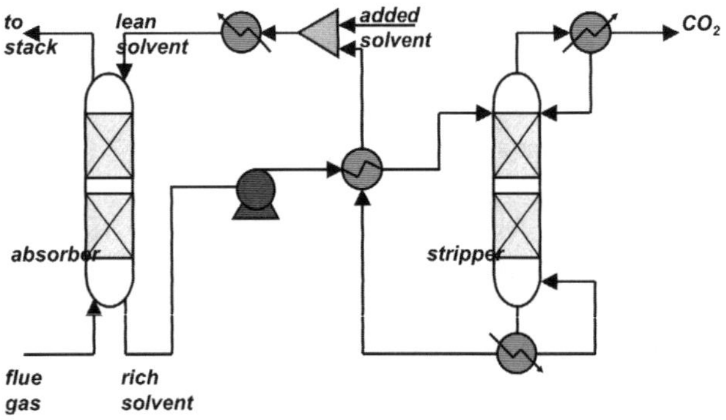

- ❑ Phase Equilibria
- ❑ Environmental
- ❑ Rheological
- ❑ **Mechanical**
- ❑ Diffusional
- ❑ **Volumetric**
- ❑ **Interfacial**
- ❑ Reactivity
- ❑ **Electrical**
- ❑ **Sensory**
- ❑ Thermal
- ❑ **Optical**
- ❑ Safety

2000-11-15-01

Molecular Knowledge Systems, Inc.

Page 22

Approach: Step 2

Three Step Approach

1. Developed Property Constraints:
   - Sources, Properties, Values
2. Generated Candidates:
   - chemicals, mixtures, structure
3. Evaluated Candidates:
   - computations, experiments

Generating Candidates: Chemicals

$-CH_3$	$-CH_2-$
$>CH-$	$-OH$
$-NH-$	$-CO-$
$-COO-$	$-O-$

Group Basis

- ❑ All Combinations
- ❑ Each Group Limited
- ❑ Specify Any Group
- ❑ Valence Feasible
- ❑ #b / 2 = #g - 1 + r

Generating Candidates: Chemicals

$-CH_3$ $-CH_2-$

$>CH-$ $-OH$

$-NH-$ $-CO-$

$-COO-$ $-O-$

Combinations of 2

- (-CH3 -CH3)
- (-CH3 -CH2-)
- (-CH3 -OH)
- . . .

Combinations of 3

- (-CH3 -CH3 -CH3)
- (-CH3 -CH3 -CH2-)
- . . .

Generating Candidates: Chemicals

Structural Constraints Only

Possible Groups	Groups in Molecule	Feasible Candidates
11	5	207
11	10	12,885
11	15	221,667
21	5	7,748
21	10	2,815,555

Approach: Step 3

Three Step Approach

1. Developed Property Constraints:
   - Sources, Properties, Values
2. Generated Candidates:
   - chemicals, mixtures, structure
3. **Evaluated Candidates:**
   - computations, experiments

2000-11-15-01

Evaluate: Computationally

Estimation Techniques

- ❑ **Group Contribution Techniques**
- ❑ **Equation Oriented Techniques**
- ❑ Force Field Techniques
- ❑ Connectivity Indices
- ❑ Parameter Fits

2000-11-15-01

Physical Property Estimation

Need to Screen Thousands of Candidates

⋮

CH3-CH2-CH2-CH2-OH

CH3-CH(CH3)-CH2-OH

CH3-CH(OH)-CH2-CH3

CH3-C(CH3)(CH3)-CH2-OH

CH3-C(OH)(CH3)-CH2-CH3

⋮

- ☐ Activity Coefficient
- ☐ Enthalpy of Fusion
- ☐ Melting Point
- ☐ Flash Point

2000-11-15-01

Group Contribution Techniques

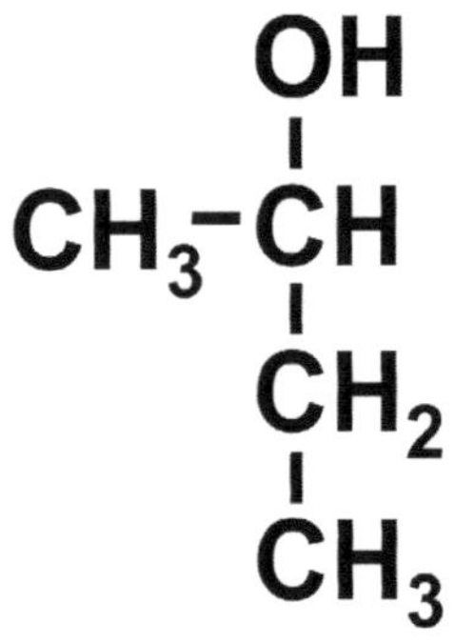

Group	ΔTb
-CH3	23.6
-CH3	23.6
-CH2-	22.9
>CH-	21.7
-OH	92.9
Tb (est)	382.8 K
Tb (lit)	372.7 K

1) Select Technique
2) Dissect Structure
3) Get Contributions
4) Insert into Model

$$Tb = 198.1 + \Sigma\Delta_i$$

2000-11-15-01

Equation Oriented Techniques

$$\ln\left(\frac{P_{vp}}{P_c}\right) = T_{br}\frac{\ln(P_c)}{1-T_{br}}\left(1-\frac{1}{T_r}\right)$$

Required Properties

Tc – Critical Temperature

Pc – Critical Pressure

Tb – Boiling point

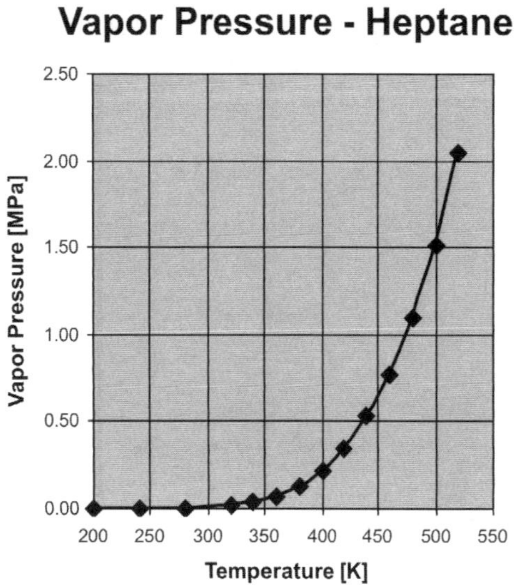

Mixture Techniques

$$\ln(\eta_m) = x_1\ln(\eta_1) + x_2\ln(\eta_2) + x_1x_2G$$

Mixture Functions

- ❑ Pure Data Inputs
- ❑ Reduces to Pure
- ❑ Molecular Forces
- ❑ Characteristics

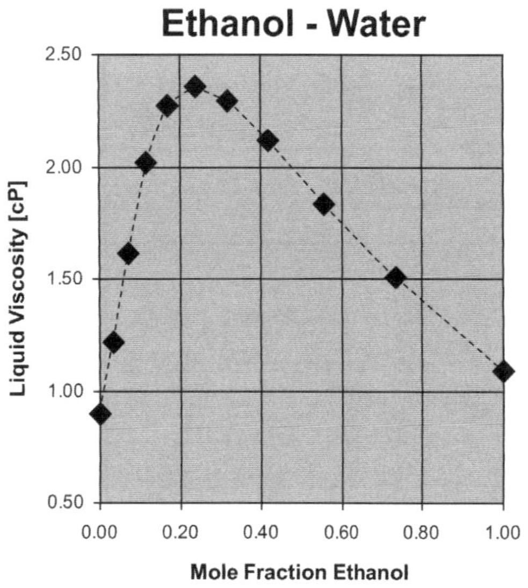

Approach

Three Step Approach

1. **Developed Property Constraints:**
   - Sources, Properties, Values
2. **Generated Candidates:**
   - chemicals, mixtures, structure
3. **Evaluated Candidates:**
   - computations, experiments

2000-11-15-01

Goal

To have you appreciate the importance of physical properties in all your decisions regarding process and product development.

What are your "gold bars"? Are they refrigerants, lubricants, solvents, adhesives, …?

Hopefully you have people working every day to replace your current products with better ones.

Your competitors sure do.

2000-11-15-01

Thank You

Questions ?

2000-11-15-01

Molecular Knowledge Systems, Inc. Page 35

Developing and adapting novel ideas for plant upgrades

Scott Siegmund, Monty Alger
Air Products

American Institute of Chemical Engineers
June 2012

General Business Objective

Profits = Sales - Costs

- Growth
- Margin %
- Fast
- *Sustainable*

AIR PRODUCTS

Segmentation – Investment / Margin

Products

Products	Existing (Process)	New (Process)
New	• New Products • New Applications • Use same process	• Product by Process • New Process Enables New Product Technology
Existing	•Core Business •Debottlenecking •Productivity	• Capacity Expansion • Leverage and extension – same

Process

AIR PRODUCTS

	RISK	Opportunity	Commitment
Process Optimization			
• *Efficiency, Cost Reduction* • *Existing Process* • *Debottleneck existing plant* • *Low cost investment*	LOW	LOW	~ 1 Yr
Process Stretch			
• *Add capacity to selected units* • *Selected Process Improvements*	LOW - MED	MED	~ 2 Yr
New Process Development			
• *New chemistry to change feedstocks* • *New flowsheet to alter investment*	HIGH	HIGH	~ 5+ Yr

Need to Balance Short-term and Long-term Efforts

The Income Statement

Revenues	$600		*Top Line*
Variable Costs		325	
Direct Materials	150		
Conversion	150		
Distribution	25		
Contribution Margin	275		
•Base Costs		200	
Executive C&B	50		
Depreciation	100		
Other SG&A	50		
Operating Margin		75	
Interest Expense		0	
Income Before Taxes		75	
Taxes		30	
Net Income		45	*Bottom Line*

Influencing the Income Statement

New Products, Delivery,

Revenues	$600		= Volume X Selling Price
Variable Costs		325	
Direct Materials	150		Short-Term Variable Cost Productivity
Conversion	150		
Distribution	25		
Contribution Margin	275		= Revenues - Variable Costs
Base Costs		200	
Executive C&B	50		
Depreciation	100		
Other SG&A	50		
Operating Margin		75	= CM - Base Costs
Interest Expense		0	
Income Before Taxes(IBT)		75	
Taxes		30	= IBT X Tax Rate (40% in this example)
Net Income		45	

Short Term: Harder to Find; Need Integrated Effort
Long Term: Need a Multi-year process plan & sustained effort

Break Even Analysis

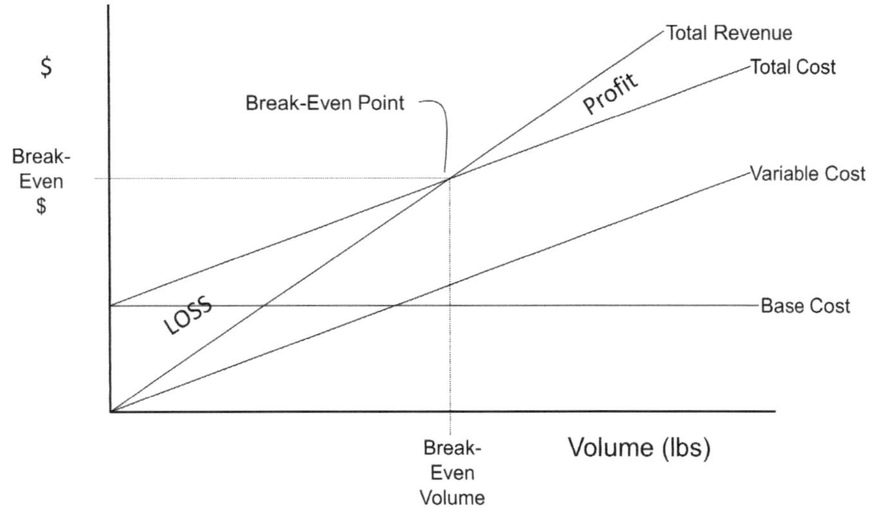

•At the Break-Even Point Net income equals 0, Contribution Margin balances base costs and Total Revenue equals Total Cost

•If the slope of the Revenue Line is less than or equal to the slope of the Variable Cost Line, profitability is impossible.

The Income Statement

Impact

		Base Case 600 MM lbs			Case 1 800 MM lbs	
		MM$	$/lb		MM$	$/lb
Revenues	$600	1.00		$800	1.00	
Variable Costs		325			399	
Direct Materials		150	0.25		200	0.25
Conversion		150	0.25		160	0.20
Distribution		25	0.04		33	0.04
Contribution Margin(%)		275 (45.8%)			407 (50.8%)	
Base Costs		200			200	
Executive C&B		50	0.08		50	0.06
Depreciation		100	0.17		100	0.13
Other SG&A		50	0.08		50	0.06
Operating Margin (%)		75 **(12.5%)**			207 **(25.8%)**	
Interest Expense		0			0	
Income Before Taxes		75			207	
Taxes		30			83	
Net Income		45			124	

Realities of Price and Cost

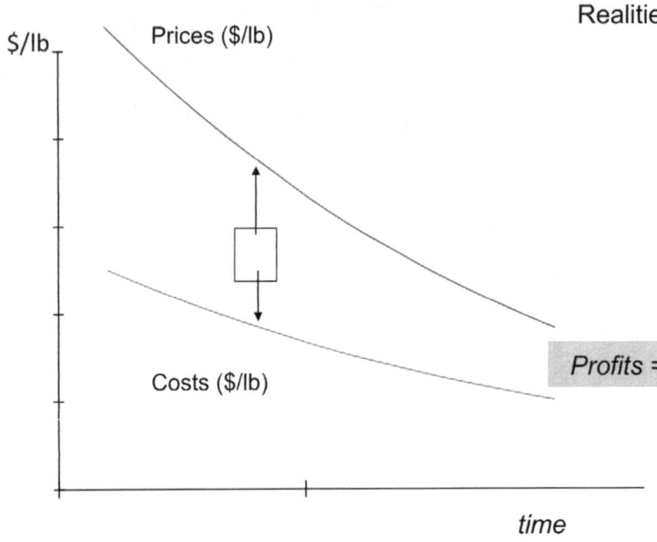

Profits = Δ x Volume

Top-line: *Growth*
- Quality
- New Products
- Product Redesign
- Services

Bottom-line: *Productivity*
- Process Optimization
- Process Stretch
- New Process Development
- Product Redesign
- Product Rationalization

For a given product over time:
- Competition and Customers will drive price down
- Costs must continue to drop as well to maximize Δ
- As Δ shrinks must focus on Volume to maintain profitability
- Otherwise, shift to new products, rationalize unprofitable ones

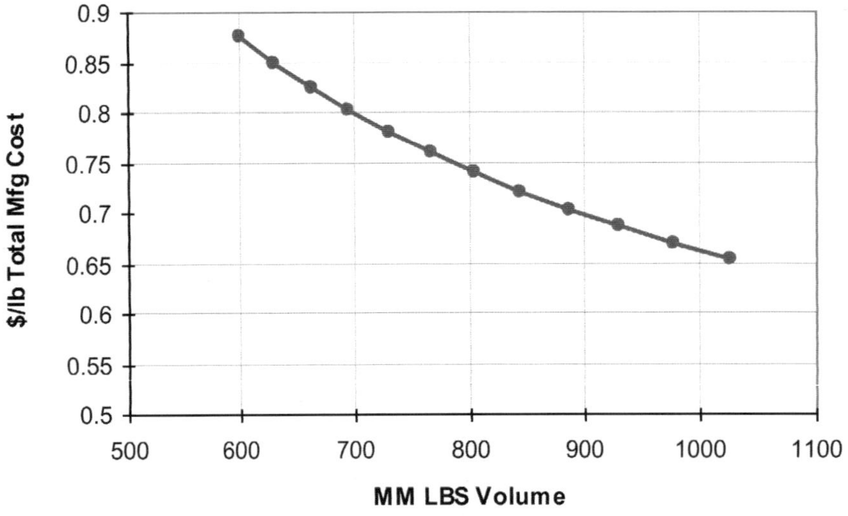

U.S. Industrial Power Costs

¢/KHW

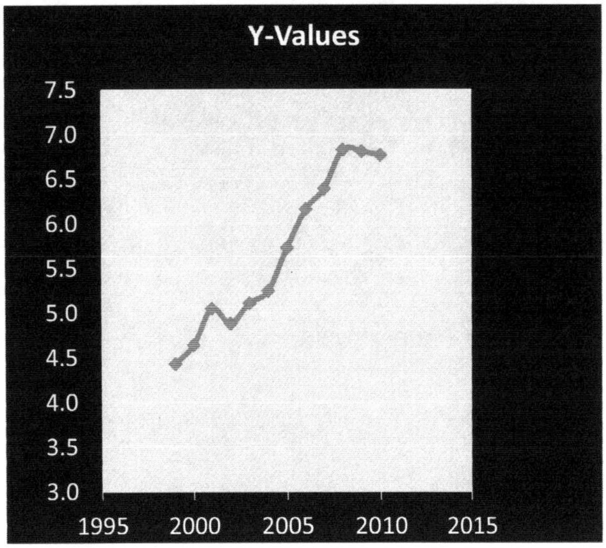

How do we promote this culture of value driven improvement inside of our companies?

First Rule

- The Business area, Operations. Engineering and R&D must communicate, really communicate.
  - Regular joint meetings
    - The Business area must present results, numbers and growth expectations.
    - The technical area must discuss new products and know why they are working on them
    - The operating group must discuss potential problems
  - Joint group CI events
  - Co-location

Example PSA to TPSA

- Air separation units require removal of CO_2 and H_2O before it enters the cryogenic portion of the plant.
- This step can often be the facility bottleneck

Front-End PSA System

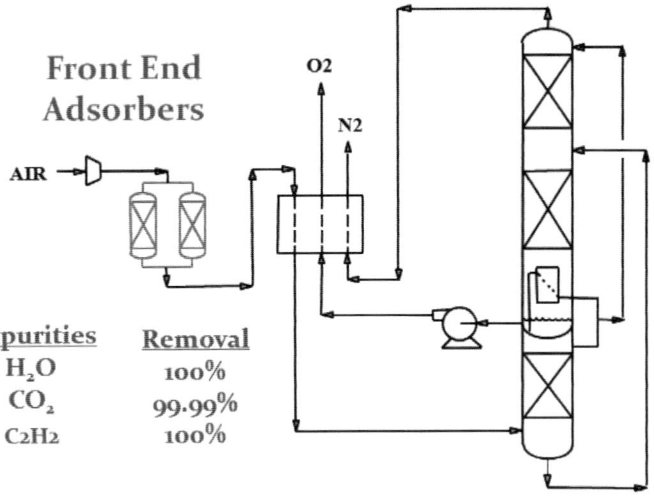

Impurities	Removal
H_2O	100%
CO_2	99.99%
C2H2	100%

How it happened

- Business area expressed an interest in getting more product out of the plant at one of the quarterly meetings.
- Operations and Technology did a plant assessment to confirm the plant bottlenecks.
- Technology develop an enhancement of the PSA system based in part on other products in the system

How it happened

Cont.

- Technology developed a pilot scale plant to test the proposed upgrade
- The Technology area along with Engineering, Operations and the Business are developed the business case and got senior management approval for the modification
- The modifications were made.

Modification (for the engineers in the audience)

- Upgrade the cleanup step to require less flow but do a better job of cleaning up the adsorbent bed.
- Did that by adding a slight amount of heat to the cleanup flow.

Case Study: PSA to TEPSA Debottleneck

AIR PRODUCTS

How did it work out?

- Results
  - 10% increase in plant production
  - 2% decrease in plant specific power.

The Income Statement

Impact

		PSA 600 MM lbs			TPSA 660 MM lbs	
		MM$	$/lb		MM$	$/lb
Revenues	$600	1.00		$660	1.00	
Variable Costs		325			335	
Direct Materials		150	0.25		147	0.244
Conversion		150	0.25		147	0.223
Distribution		25	0.04		27	0.04
Contribution Margin(%)		275 (45.8%)			339 (51.4%)	
Base Costs		200			201	
Executive C&B		50	0.08		50	0.06
Depreciation		100	0.17		101	0.13
Other SG&A		50	0.08		50	0.06
Operating Margin (%)		75 **(12.5%)**			138 **(20.9%)**	
Interest Expense		0			0	
Income Before Taxes		75			124	
Taxes		30			55	
Net Income		45			83	

Global Challenges

- Pollution
- Population Growth
- Energy, Water
- Resource Shortages

AIR PRODUCTS

Sustainability

Source: Professor Roland Clift, Centre for Environmental Strategy (CES), University of Surrey; www.surrey.ac.uk

AIR PRODUCTS

Thank You!

Product & Process Design Options

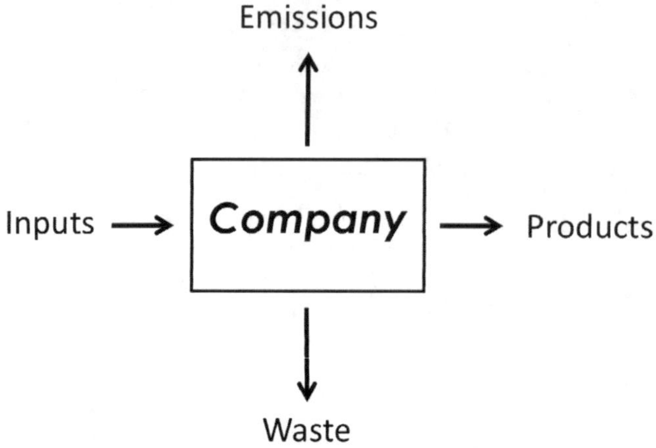

Sustainability: "People, Planet, Profits"

Transparency: Demands are increasing

Global Reporting Initiative

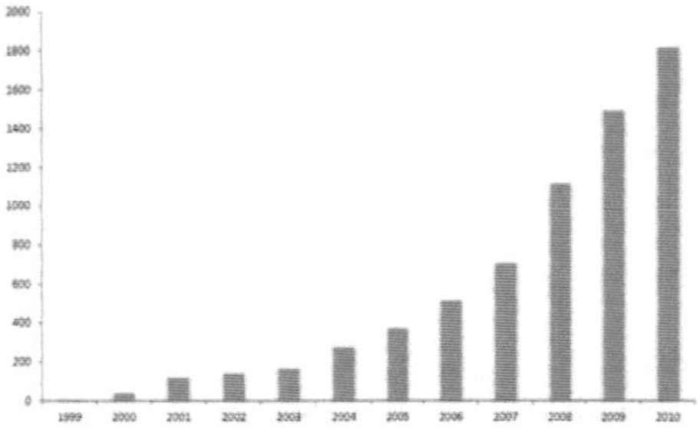

Year	Growth
2010	↑ 22%
2009	↑ 34%
2008	↑ 58%
2007	↑ 37%
2006	↑ 38%
2005	↑ 36%
2004	↑ 65%
2003	↑ 20%
2002	↑ 14%
2001	↑ 177%
2000	↑ 300%
1999	base year

Source: Global Reporting Initiative
GRI Sustainability Reporting Statistics
Publication year 2010
By Report Services

http://www.globalreporting.org

Carbon Disclosure Project

Sector	Company	Disclosure score
Information Technology	Cisco Systems	98
	Accenture	93
	SAP	96
	Sony Corporation	94
	Samsung Electronics	94
Materials	Lafarge	96
	Dow Chemical	95
	VALE	93
	Praxair	93
	BASF	93
	Air Products & Chemicals	92
	Israel Chemicals	90
Telecommunications	Telefonica	90
Utilities	Fortum Oyj	97
	Centrica	96
	PG&E	92

Key:

Companies highlighted in green are those that have been in the Global 500 CDLI for three consecutive years (2009-2011)
Companies highlighted in orange have moved into the Global 500 CDLI index this year

http://www.cdproject.net/CDPResults/CDP-G500-2011-Report.pdf

Ref: 7-7-1EEGW

Product & Process Design Innovation

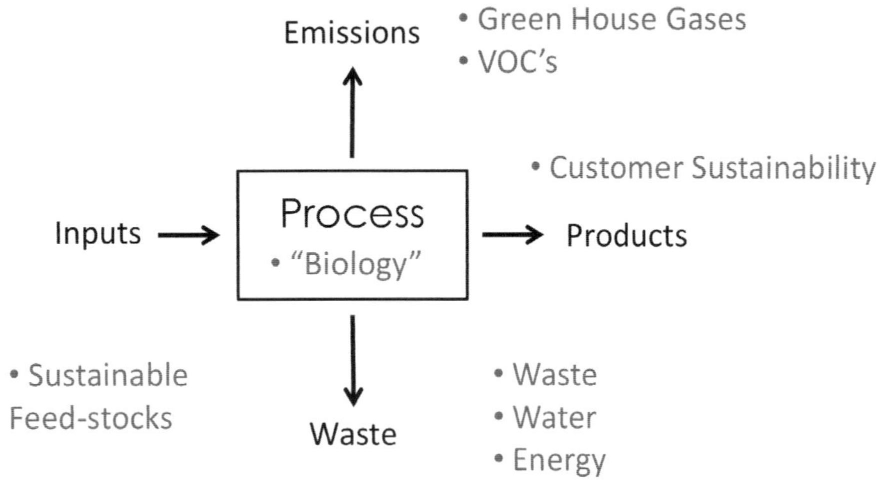

Multiple New Sustainability Requirements & Transparency

Typical Project Cash Flow / IRR

Income Statement
Show variable, fixed costs
Highlight Investment in the depreciation as major cost

Scale with types of $ investment / $ of sales for capital intensity

Operating Margin

Show a cash flow related to examplee financials to the left side.

Project Segmentation – Investment / Margin

Investment $	Low	High
High	•Major Capacity Expansion •New Capacity	• Product by Process • New Process Enables New Product Technology
Low	•Core Business •Debottlenecking •Productivity	•New Applications •New Product Development
		Margin %

Options

Revenues

Price X - Price / Unit Sold (e.g. $/lb)

Volume - Amount of Product Sold (MM LBS)

Costs

Variable Costs - Change with Volume

Base Costs - Generally Fixed

Profits

	RISK	Opportunity	Commitment
Process Optimization • *Efficiency, Cost Reduction* • *Existing Process* • *Debottleneck existing plant* • *Low cost investment*	LOW	LOW	~ 1 Yr
Process Stretch • *Add capacity to selected units* • *Selected Process Improvements*	LOW - MED	MED	~ 2 Yr
New Process Development • *New chemistry to change feedstocks* • *New flowsheet to alter investment*	HIGH	HIGH	~ 5+ Yr

Need to Balance Short-term and Long-term Efforts

Project Segmentation

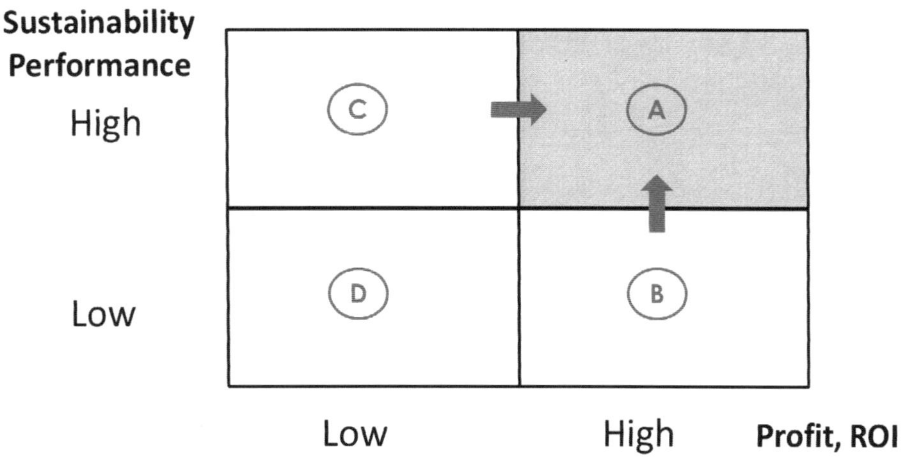

Some Options:

- **Markets, Applications**
- **Products & Processes**
- **First time / Reinvestment**

Which is More "Sustainable" ?

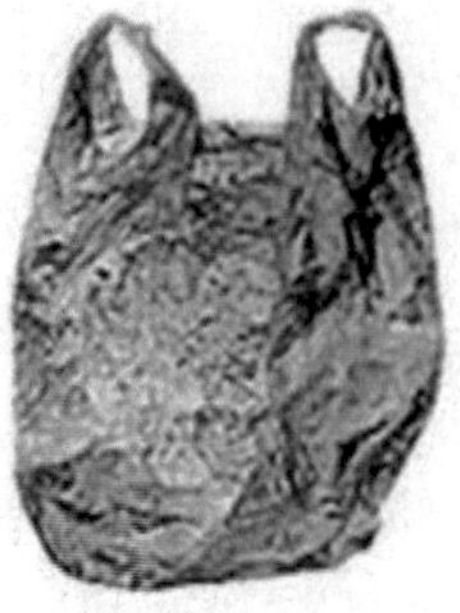

Paper Bag (Recycled Content)

Plastic Bag

Footprint for Paper and Plastic Bags

Impact Summary of Various Bag Types (Carrying Capacity Equivalent to 1000 Paper Bags)	**Paper (30% Recycled Fiber)**	**Polyethylene**
Total Energy Usage (MJ)	2,622	763
Fossil Fuel Use (kg)	23.2	14.9
Municipal Solid Waste (kg)	33.9	7.0
Greenhouse Gas Emissions (CO_2 Equiv. Tons)	0.08	0.04
Fresh Water Usage (Gal)	1,004	58
Landfill Half-Life	Short	?

Sources:
- **"Bag fact sheet"**, American Chemistry Council
- **" Life Cycle Assessment for Three Types of Grocery Bags - Recyclable Plastic; Compostable, Biodegradable Plastic; and Recycled, Recyclable Paper "**, Boustead Consulting & Associates Ltd

Mike Schultz
VP Engineering
LanzaTech

AIChE Process Development Symposium
King of Prussia, PA
June 7, 2012

LanzaTech

Feeding the Funnel

LanzaTech

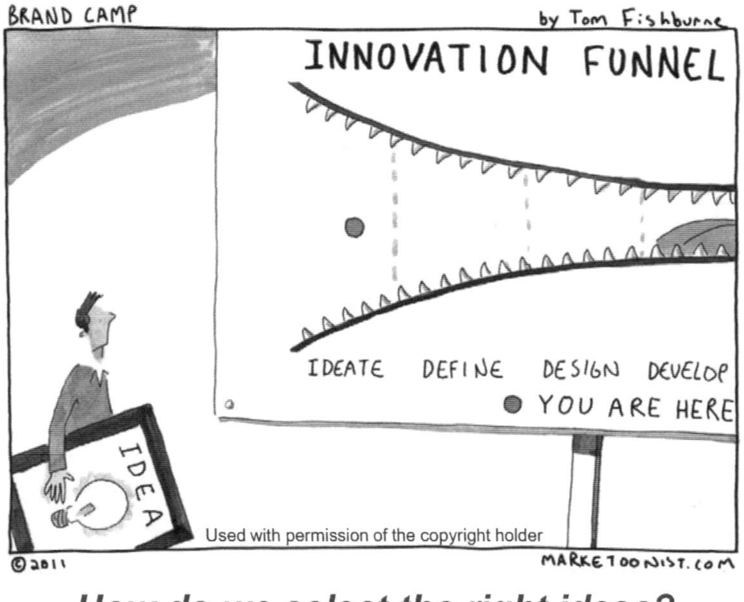

How do we select the right ideas?
How do we develop these ideas?

Guiding the Innovation Process

LanzaTech

From the Blog 'Innovate on Purpose', Jeffrey Phillips

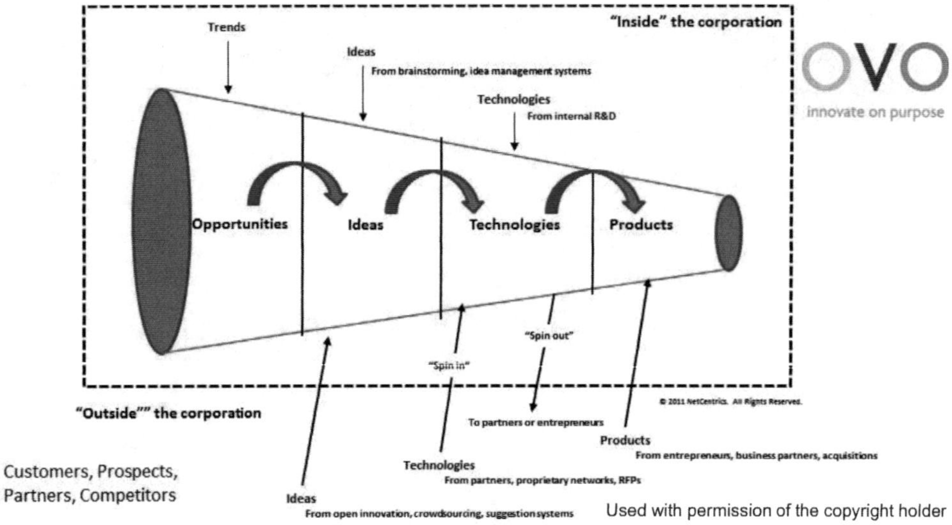

Used with permission of the copyright holder

Efficient use of $ and Resources—Maximize return on Investment

Getting the Right Ideas Into the Funnel

LanzaTech

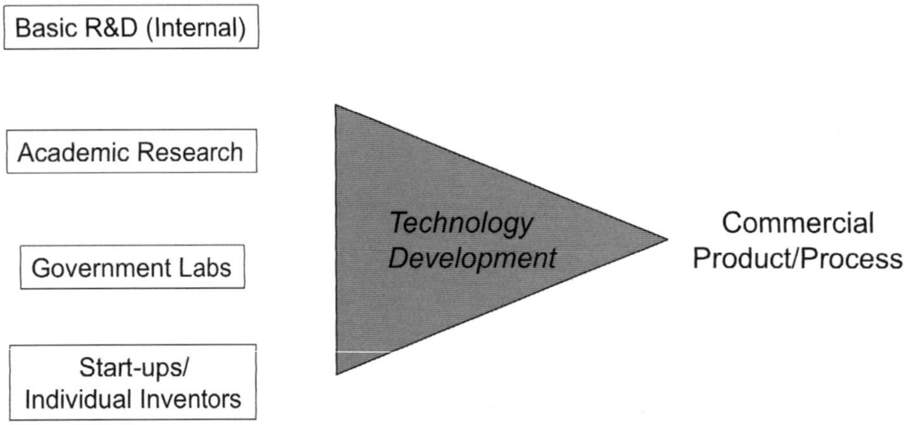

Technology Scanning Can Identify External Opportunities

Match Market/Customer Needs with Potential Solutions

Early Stage Process Design

LanzaTech

- **Use early stage process design to identify and select concepts**
  - **Does the idea fit in with our commercial objectives?**
  - **Does the idea follow the basic laws of science/engineering/economics?**
  - **Chemical engineering 101: what is my overall mass and energy balance, and are the products worth more than the feeds (Douglas Level 1)**
- **Establish targets: technical, economic, and life cycle**
  - **What experiments/data do I need?**
  - **Focus on the steps with the greatest impact to economics and LCA**

Things to Watch Out For

LanzaTech

Doing the Wrong Analysis ⟹
- Shutting the Project Down Too Early
- Letting the Project go to Far

Doing too Much Analysis ⟹
- Getting caught up in details
- Not moving fast enough
- Not able to adapt

Not Understanding the Problem, or Market, or Data ⟹
- Missing the opportunity
- Over/underestimating the opportunity

6

Bringing Process Engineering Into Early Stage Development

LanzaTech

Requires Science/Engineering Interaction

Decisions
- Reaction Conditions
- Solvent type/amount
- Catalyst type/amount
- Co-feed type/amount
- Separation process/conditions
- ....

Process Design
- Product yield/purity
- Byproduct type/amount
- Waste stream type/amount
- Equipment type/size
- Utility type/magnitude
- ...

Economics/
Sustainability

Should be More Collaborative, Less Iterative

- **Decisions drive the process design, but the economics and sustainability assessment then inform the decision process**
- **Need fairly accurate estimates of process design, economics and impact on sustainability, but done with a minimal amount of time, effort, and money**

Examples

LanzaTech

- **Example #1: Taking an idea from conception to commercialization**
- **Example #2: Using conceptual process design to rule out a potential collaboration**
- **Example #3: Incorporating LCA criteria to inform the process design**

Example #1: Dividing Wall Column Commercialization

LanzaTech

- **Commercial Challenge:**
  - **Provide solution to reduce capital cost**
  - **Need it quickly**
- **Technical expert identifies DWC as a potential solution**
- **Small team convened to evaluate potential advantage**

Dividing Wall Column Overview

LanzaTech

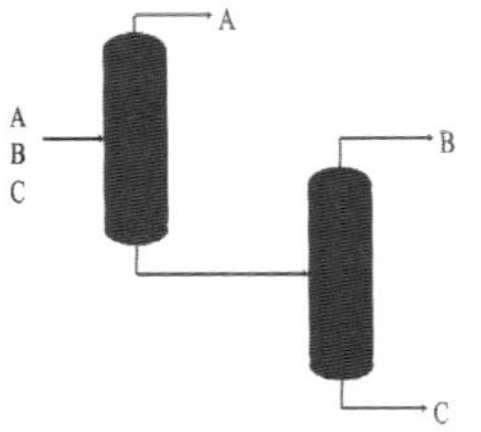

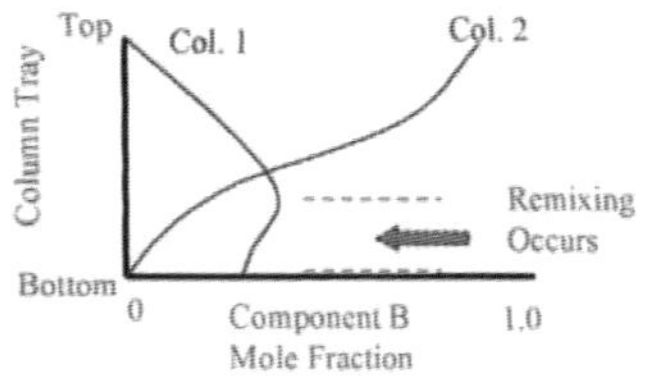

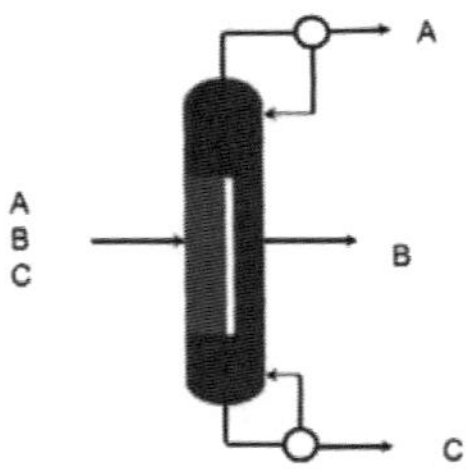

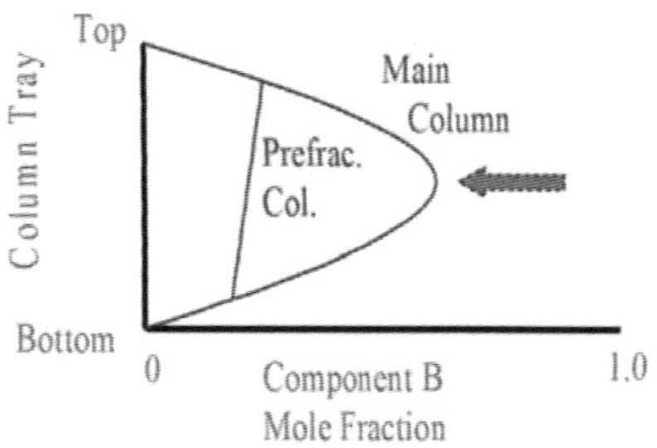

- **DWC provides more efficient separation**
- **Reduces energy *and* capital: there is no tradeoff!**

Example #1 Continued

- **Step 1:**
  - **Develop preliminary design concept**
  - **Evaluate the potential benefit**
  - **Refine the concept**
- **Step 2: Key questions identified**
  - **Process control**
  - **Process operation**
  - **Mechanical integrity**
  - **Tray design**
- **Step 3: Key stakeholders engaged--engineering, operation, commercial**
- **Step 4: Startup--work with startup crew to identify data needs, tests to run during startup**

Example #2: Evaluating a Potential Collaboration

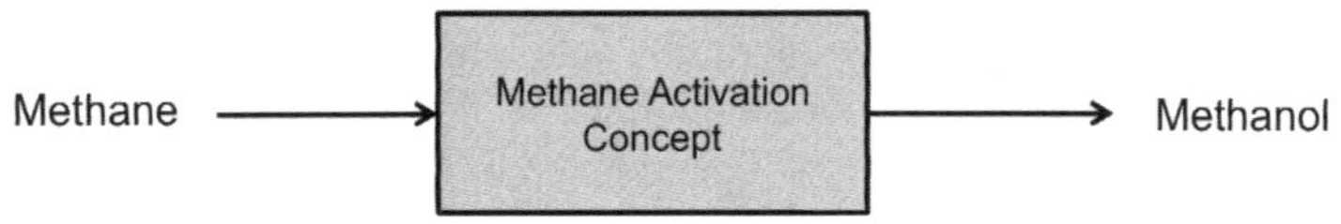

✓ Commercial Goals: Develop NG Portfolio
✓ Market Need: Monetize small/stranded NG fields
✓ Preliminary Evaluation:
- Material balance completed
- Key technical challenges identified
- Competing technology model (MeOH) developed

Example #2: Preliminary Evaluation

LanzaTech

Preliminary Process Model

- Overall Material Balance and Preliminary Cost Estimates
- Where do I Need More Data?
- Which Process Steps Require More Detail?

Commercial Evaluation

- Market Drivers
- Where are the Opportunities?

Key Technical Challenges

- Which Process Steps are Critical
- What can be done In-house vs. Externally?

Example #2: Deeper Dive

LanzaTech

- **Collaboration Discussions: IP Valuation, Potential Partners**
- **Key conversion step: SO_2 oxidation**
  - **Initially thought to be trivial, so not evaluated in much detail**
  - **Option 1: Required an H_2SO_4 plant several times larger than largest commercial**
  - **Option 2: Develop new SO_2 oxidation technology**

These Challenges Were Sufficient Such That Collaboration Was Not Pursued

Example #3: Incorporating LCA to Early Stage Process Design

LanzaTech

- Methods have been developed for early stage conceptual process design
  - Hierarchical—focus on key economic decisions early, add more detail later
  - Use methods to estimate CAPEX and OPEX without detailed process design
  - Close interaction with early stage chemistry/biology
    - Use experimental results to develop flowsheet alternatives and estimate economics
    - Use process design results to identify most economic flowscheme alternatives—propose experiments to evaluate as necessary
    - Identify gaps in technical knowledge—propose experiments
- Extension to incorporate sustainability/life cycle assessment

Example #3 Continued

LanzaTech

- What are the LCA drivers?
  - Regulatory
  - Carbon Credits
  - Mandates
  - Corporate Mission
- Establish LCA targets
  - E.g. X % reduction in CO2 emissions
  - Minimize water loss
  - ...
- Process models should include factors for LCA
  - Focus on those intrinsic to the process: e.g. energy requirements and sources
  - Adds new element to capital/energy trade-off

Example #3 Continued

LanzaTech

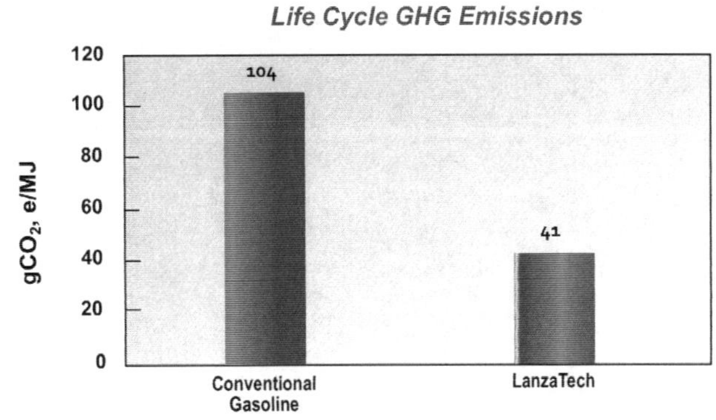

Fermentation of Steel Mill Offgases to Ethanol

- **Target: Reduction in CO2 emissions relative to gasoline**
- **Assessment:**
  - **Impact of increased capital on energy needs**
  - **Tradeoffs in utilities: e.g. compression vs. steam**

17

Conclusions

LanzaTech

- **Early stage process design provides a tool to help select the right ideas for technology development**
- **Early stage process design can then be used to guide the innovative process towards commercial products and processes**
- **The key is balancing rigor and detail vs. time and complexity**

Making the most of R&D resources to deliver commercial results

PARALLEL UNIVERSES: The Science of Learning from Others

Jack Hipple
Innovation-TRIZ
Tampa, FL
www.innovation-triz.com
jwhinnovator@earthlink.net

Issues

- Scale up challenges
  - Technology
  - Globalization
  - ***<u>Not reinventing the wheel</u>*** **(time and money are precious!)**
- Parallel universes we can learn from
- Translating to our situation
- Examples and suggestions

Psychological Inertia

- We frame the problem or challenge we have in terms we use and those around us can understand
  - Technical jargon (makes us feel special, right?)
  - Acronyms, shortened chemical names
    - Only "we" understand
    - "GUI"
- We don't believe anyone else has already solved a similar problem
  - If someone has, why would we be working on it?

Psychological Inertia *<u>is Important!</u>*

- Process control
- Product specifications
- Standards
- Inspection procedures
- Flight plans
- Recipes and formulas
  - Latexes
  - Gasoline blends
  - Pharmaceuticals
  - Ag chemicals

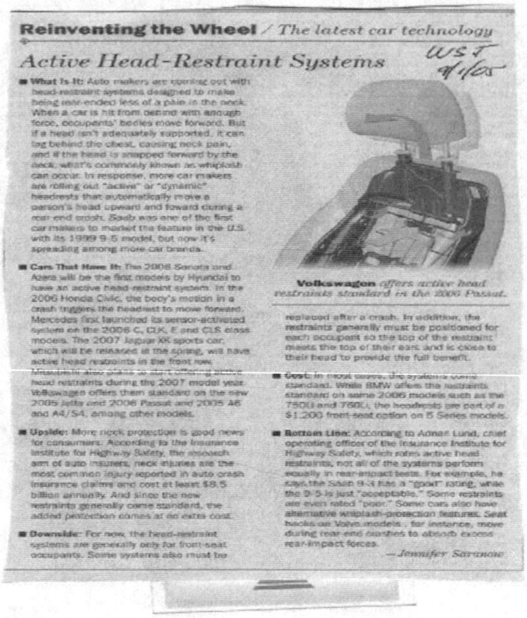

Reinventing the Wheel / *The latest car technology*

Active Head-Restraint Systems

WSJ 9/1/05

■ **What Is It:** Auto makers are coming out with head-restraint systems designed to make being rear-ended less of a pain in the neck. When a car is hit from behind with enough force, occupants' bodies move forward. But if a head isn't adequately supported, it can lag behind the chest, causing neck pain, and if the head is snapped forward by the neck, what's commonly known as whiplash can occur. In response, more car makers are rolling out "active" or "dynamic" headrests that automatically move a person's head upward and foward during a rear-end crash. Saab was one of the first car makers to market the feature in the U.S. with its 1999 9-5 model, but now it's spreading among more car brands.

■ **Cars That Have It:** The 2006 Sonata and Azera will be the first models by Hyundai to have an active head-restraint system. In the 2006 Honda Civic, the body's motion in a crash triggers the headrest to move forward. Mercedes first launched its sensor-activated system on the 2006 C, CLK, E and CLS class models. The 2007 Jaguar XK sports car, which will be released in the spring, will have active head restraints in the front row. Mitsubishi also plans to start offering active head restraints during the 2007 model year. Volkswagen offers them standard on the new 2005 Jetta and 2006 Passat and 2005 A6 and A4/S4, among other models.

■ **Upside:** More neck protection is good news for consumers. According to the Insurance Institute for Highway Safety, the research arm of auto insurers, neck injuries are the most common injury reported in auto crash insurance claims and cost at least $8.5 billion annually. And since the new restraints generally come standard, the added protection comes at no extra cost.

■ **Downside:** For now, the head-restraint systems are generally only for front-seat occupants. Some systems also must be replaced after a crash. In addition, the restraints generally must be positioned for each occupant so the top of the restraint meets the top of their ears and is close to their head to provide the full benefit.

Volkswagen *offers active head restraints standard in the 2006 Passat.*

■ **Cost:** In most cases, the systems come standard. While BMW offers the restraints standard on some 2006 models such as the 750Li and 760Li, the headrests are part of a $1,200 front-seat option on 5 Series models.

■ **Bottom Line:** According to Adrian Lund, chief operating officer of the Insurance Institute for Highway Safety, which rates active head restraints, not all of the systems perform equally in rear-impact tests. For example, he says the Saab 9-3 has a "good" rating, while the 9-5 is just "acceptable." Some restraints are even rated "poor." Some cars also have alternative whiplash-protection features. Seat backs on Volvo models, for instance, move during rear-end crashes to absorb excess rear-impact forces.

—Jennifer Saranow

What is the Most General Way We Can Describe This Problem?

What Parallel Universes Might Have a Similar Type of Problem?

Google Search

- "Improve auto headrests"
  - 650,000 hits
- "Prevent harm from moving object suddenly stopping"
  - ***<u>44,000,000 hits</u>***

Dissolving Stent: Polylactic Acid

ABT Reports Promising 3-Year Data on Bioabsorbable Stent

[These data come from a single-arm proof of concept study appropriately called ABSORB; the concept of a bioabsorbable drug-eluting stent appears to have been proven insofar as the stents were fully absorbed and the clinical outcomes were impressive to the extent that one can make such a determination in a single-arm study. ABT reported two-year data from the first phase of this study in early 2009.... and the three-year data contain no new AE's or safety signals.

To build upon the promising results of the ABSORB trial, Abbott is initiating a large-scale trial called ABSORB EXTEND, which will enroll approximately 1,000 patients from up to 100 centers in Europe, Asia Pacific, Canada and Latin America. ABSORB EXTEND is a single-arm study designed to further evaluate the performance of Abbott's proprietary fully bioabsorbable stent technology. The study will enroll patients with more complex coronary artery disease and is slated to begin enrolling before the end of the year.

"***Abbott's bioabsorbable stent has the potential to be a major breakthrough for coronary artery disease patients. The data show that patients continue to do well three years after treatment with the bioabsorbable coronary stent***," said Patrick W. Serruys, M.D., Ph.D., professor of interventional cardiology at the Thoraxcentre, Erasmus University Hospital, Rotterdam, the Netherlands, and principal investigator for the ABSORB trial. "The strong results confirm my belief that bioabsorbable technology is the next revolution in interventional cardiology."

Abbott is the only company with three-year clinical data evaluating the safety and performance of a fully bioabsorbable drug eluting coronary stent. Abbott's bioabsorbable everolimus eluting coronary scaffold is made of ***polylactide***, a proven biocompatible material that is commonly used in medical implants such as absorbable sutures. As with a metallic coronary stent, Abbott's bioabsorbable technology is designed to restore blood flow by propping open a clogged vessel, and to provide support until the blood vessel heals. Unlike a metallic stent, however, a bioabsorbable scaffold is designed to be slowly metabolized by the body and is completely absorbed over time.

Polylactic Acid Garbage Bags (Several years before Abbott)

Think about this question from a senior scientist at Abbott...

Hey, boss, can you approve my expense account for going to a garbage bag association meeting?

"The Pumps and Pipes Symposium"

- A Reynolds number is a Reynolds number...
- A friction factor is a friction factor....
- Viscosity is viscosity...
- Density is density....
- An orifice plate is an orifice plate...
- The Bernoulli equation always applies

The Boeing Corporation

What's the Best Way to Get Seat Assemblies Hoisted Up Into the Body of a 757 Jet?

Hoists and Chains? (And Whips?)

©Innovation-TRIZ, 2012

Lean Times: With Airbus on Its Tail , Boeing Is Rethinking How It Builds Planes Old Hay Loaders, New Resins Play Roles as Firm Tries To Build Faster, Cheaper

Hitting a `Rescue Me' Button
By J. Lynn Lunsford

09/05/2001 The Wall Street Journal (Copyright (c) 2001, Dow Jones & Company, Inc.)

RENTON, Wash. -- Not far from the steady blatt-blatt of the rivet guns on its 757 assembly line just outside Seattle sits what Boeing Co. calls its moonshine shop: The people here distill work-saving ideas into contraptions that make it easier to build jets.
Consider the hay loader next to an almost-completed 757. Normally, this cross between a ladder and a metal-spiked conveyor belt would be dumping bales of hay onto waiting trucks. ***But to veteran mechanic Robert Harms, the hay loader is the perfect way to get bulky passenger seats from the factory floor up 13 feet to the door of a plane without having to use an overhead crane. "It might look funny, but when you see it work, you wonder why we didn't do it this way all along," he says.***

Hay Loader from the 1600's…

n-TRIZ, 2012

Hay Loader from the 1800's…

RIZ, 2012

Boeing's "Invention": Hay Loaders...

©Innovation-TRIZ, 2012

Ball Bearing Quality Inspection

You are in the ball bearing business and want some new ideas for monitoring and controlling quality and uniformity (hardness)—where would you look?

Now: Quarantine, statistically sample, slice, examine metallurgy, ship

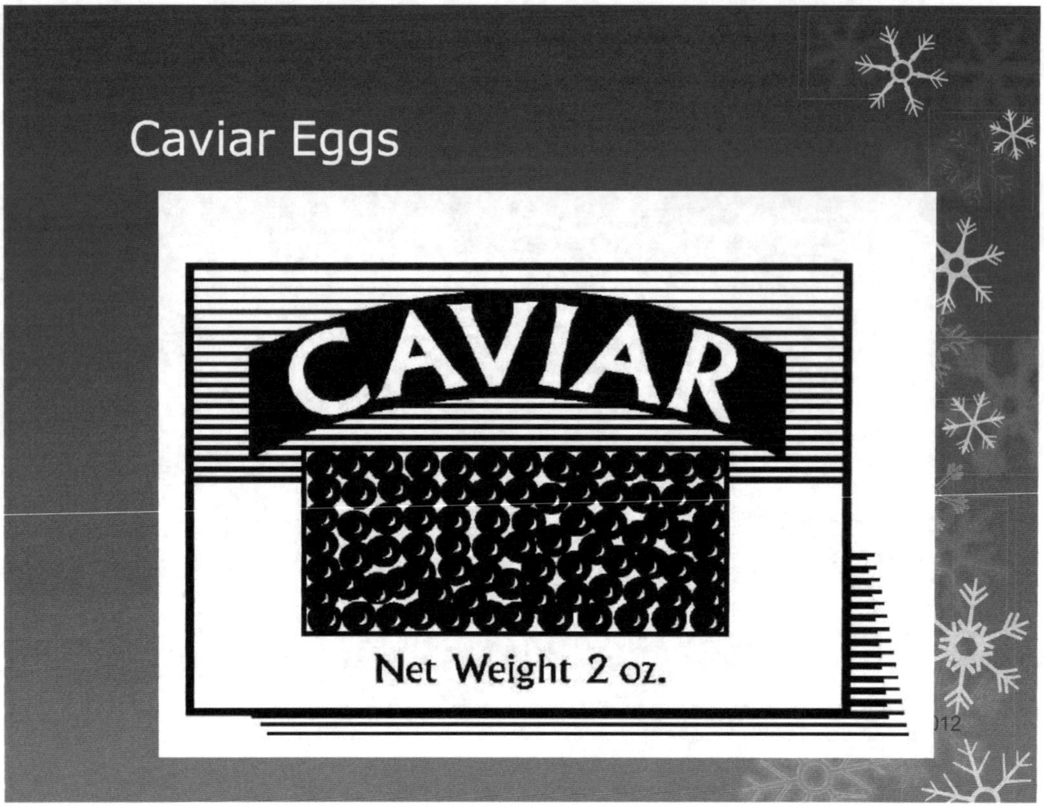
Caviar Eggs
CAVIAR
Net Weight 2 oz.

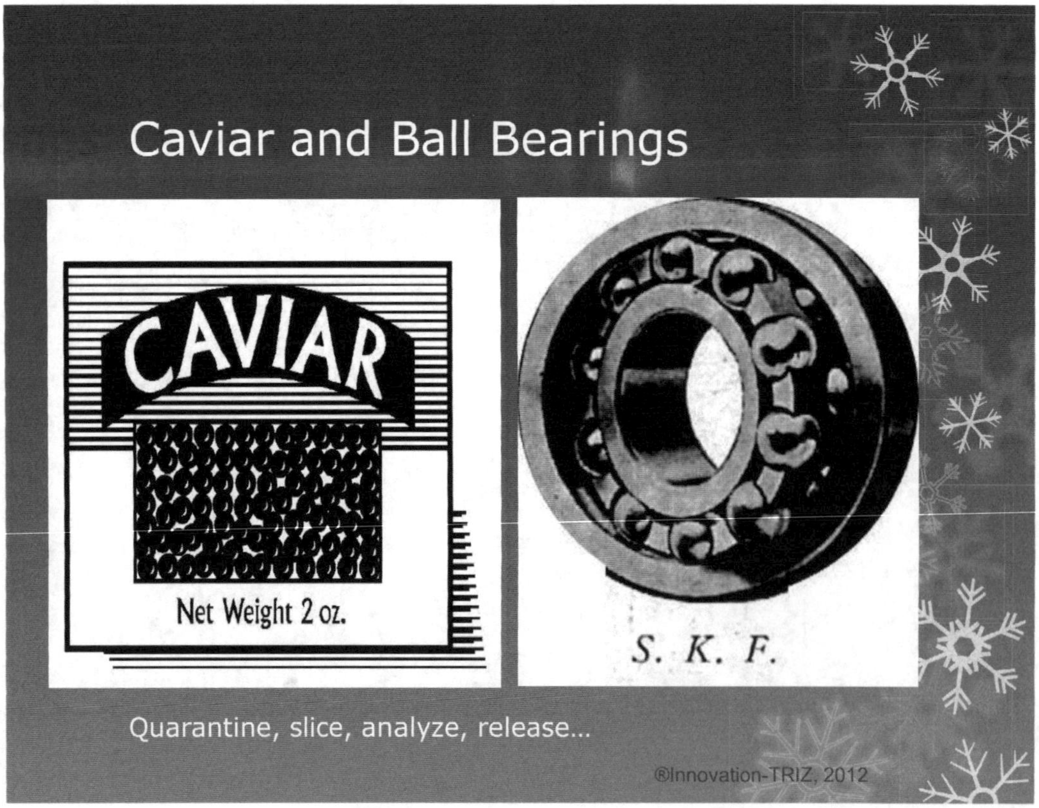
Caviar and Ball Bearings
CAVIAR
Net Weight 2 oz.
S. K. F.
Quarantine, slice, analyze, release...
©Innovation-TRIZ, 2012

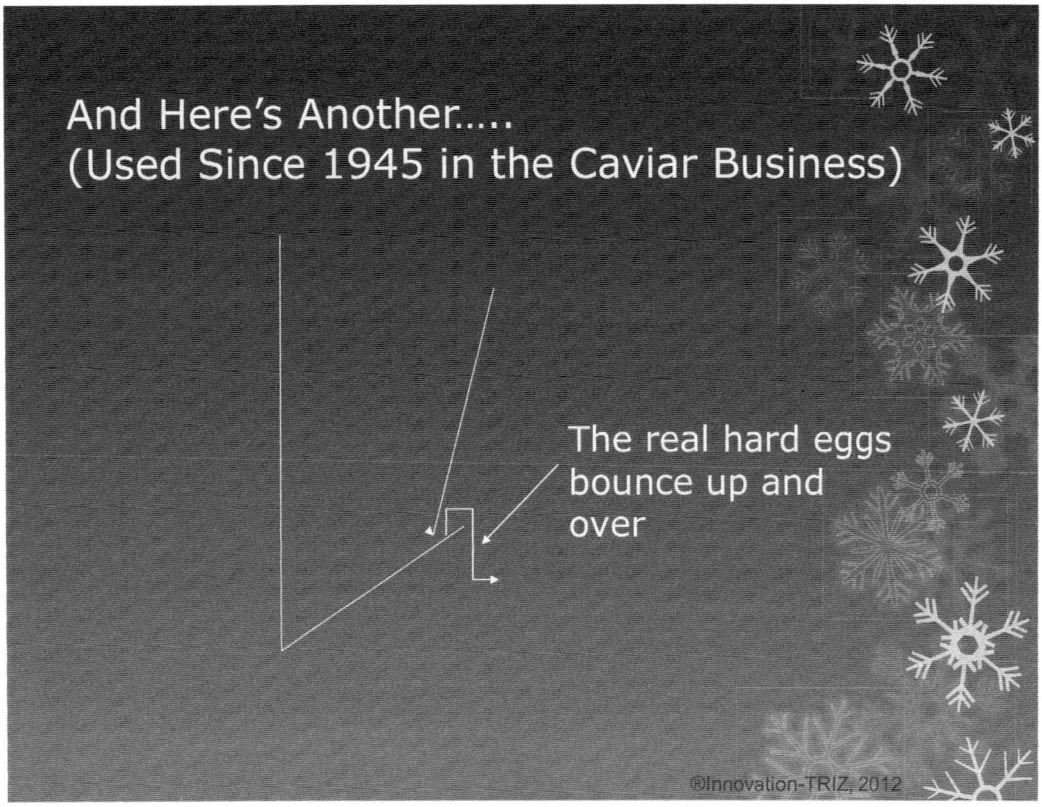
And Here's Another.....
(Used Since 1945 in the Caviar Business)
The real hard eggs bounce up and over
®Innovation-TRIZ, 2012

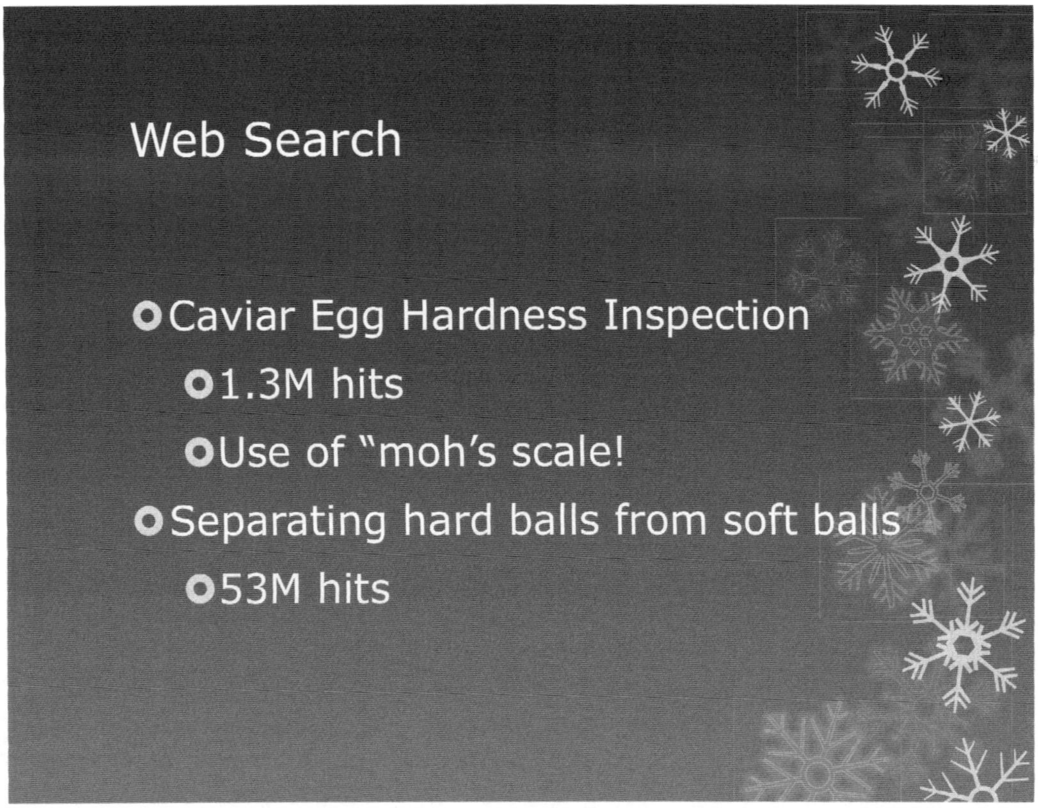
Web Search
Caviar Egg Hardness Inspection
1.3M hits
Use of "moh's scale!
Separating hard balls from soft balls
53M hits

Psychological Inertia

"The problem is never how get new, initiative thoughts into your mind, ***<u>but how to get the old ones out</u>***"

--Dee Hock, (Former) Innovation Manager, Citibank

The (Re)Discovery of the Force of Gravity

Exercise: Your Parallels
"Mommy or Daddy, What Do You Do?

- My "technology"
- Parallel universes

Who Else Needs to Perform the ***Function***
That I Need to Accomplish?

Major Challenge in Ethanol: Milling

Who else has to "mill"?

Web Search

- "Corn based alcohol plants"
  - 4.5 M hits
- "Milling"
  - 46 M hits
  - Where else is "milling" done?
  - How is it done?
- Could a continuous milling machine, using chemical addition, grind grain?
- Has a milling machine company ever talked to a grain based alcohol company?
- Has a grain based alcohol company ever been to a machine milling convention or trade show?

American Milling

- Flour Milling
- Oat Milling
- Rye Milling
- Mill Stone Milling
- Bio Energy / Bio Fuel
- Wood Pelleting Process
- Seed/Pulses/Lentil
- Feed Milling
- Hulling Milling

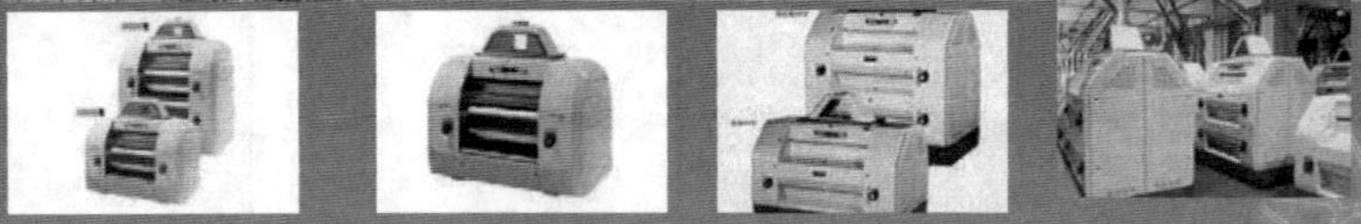

Bridgeport Milling

- Industries We Serve
- Over the last 100 years Bridgeport has been at the forefront of machine tool technological developments. Our products, combined with our unwavering commitment to helping our customers improve their productivity and business performance have contributed to our large, and loyal, customer base. We are confident that our advanced manufacturing technology will not only improve your competitive edge it will also transform your business performance. The breadth and depth of our product solutions are unmatched in the industry, enabling us to support a variety of market applications, including ***aerospace, automotive, medical, energy, construction, agricultural equipment, mold, tool and die, and more***.

How to "Translate" and "Learn from Others"

- Describe your problem to a 10 yr. old
  - Mommy or Daddy, what do you do?
  - Do you say "we try to understand the kinetics of reactive chemicals" or "we try to find out how to not make things go boom"
  - Use that language in a Google search!
    - 8 M hits
    - 46 M hits
- Think ***function***, not process

More Examples for Thought

- Separate solid from liquid....NOT filtration or crystallization
  - 73M vs. 28M
- Combining things....NOT mixing
  - 73M vs. 33M
- Transfer heat...NOT heating or cooling
  - 700M vs. 327M
- Separate liquids with different boiling points vs. distillation
  - 2.5M vs. 65M
- Move materials...NOT shipping or logistics
  - 432M vs. 277M
- Store information...NOT "the cloud"
  - 1.7B vs. 670M

Summary

- You are not alone in the universe
  - You may ***think*** you are
- Stop using acronyms and fancy words when thinking about new process development
  - Describe to a 10 yr. old
- Think about the ***function*** you need
  - Who else tries to do the same ***function?***
  - Use general terms in Google searches

Questions?
Comments?
Experiences?

Glad to talk anytime, especially if you have examples!

From Process Technology to Commercial Scale Production

Mark Shmorhun
Director Technology Transfer
Myriant Corporation
mshmorhun@myriant.com
www.myriant.com

2012 AICHE Process Development Symposium

Key Points for Successful Project Execution

In this presentation a Project is defined as follows:

Delivery of an operating dry mill ethanol plant on a greenfield site originating from a technology package.

Key Points

- Managing interfaces between disciplines and stakeholders is key to success
- Careful, detailed scope definition at those interfaces is essential
- A disciplined, detail oriented, highly organized Project Manager within each stakeholder group is critical

2

Ethanol Industry Background (2006-2008)

- **Delta-T Corporation, an ethanol technology company, licensed technology for approximately 30 new dry mill ethanol plants**

- **2006-2009 Ethanol Industry**
  - Development frenzy
  - Handful of technology providers (3-5 major players)
  - Extraordinary growth in production capacity fueled by RFS (Renewable Fuel Standard)

- **Highly Diverse Stakeholders & Interests**
  - Investors / Owners
  - Banks
  - Technology Providers
  - Engineering Firms
  - EPC Contractors
  - Plant Management

3

US Ethanol Industry Growth

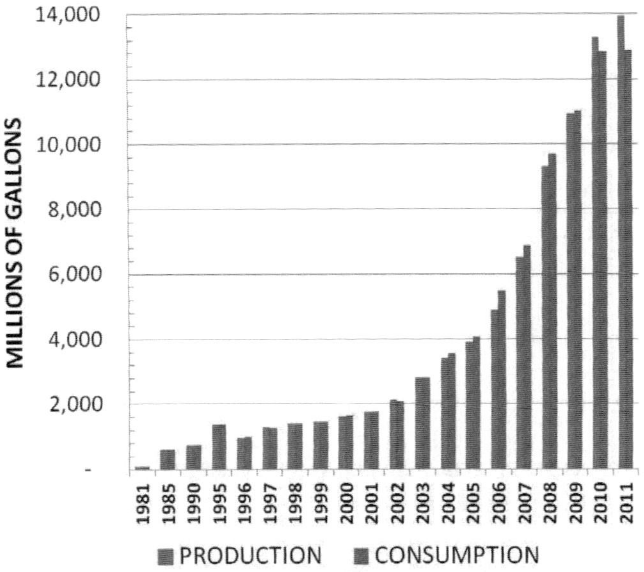

2005
~ 100 plants
2006
33 groundbreakings
2012
218 plants
$10 – 15 Billion invested

4

Project Development – Many Steps – Many Interfaces

- **Investor group selects technology provider & EPC**
- **Site Selection**
- **Basic Design Package (Schedule A Package)**
- **Cost estimation**
- **Permits**
- **Financing**
- **Detailed engineering**
- **Construction**
- **Commissioning & Startup**
- **Training**

5

Role of the Technology Provider

- **Support Many Stakeholders**
  - Investors, Financial Institutions, Plant Management
  - Regulatory Authorities
  - Engineer, Contractor, Subcontractors
  - Equipment Vendors
  - Utilities, Logistics, Raw Material Suppliers
- **Technology Package / Schedule A Package**
  - Material and Energy Balance
  - P&ID's
  - Layout Drawings
  - Lists – equipment, line, valve, instrument

6

Front End Loading (FEL)*

Front-end loading (FEL), also referred to as pre-project planning (PPP), front-end engineering design (FEED), feasibility analysis, conceptual planning, programming/schematic design and early project planning, is the process for conceptual development of projects in processing industries.

FEL-1		FEL-2		FEL-3
Material Balance	GATE	Prelim Equipment Design	GATE	Purchase Ready Major Equipment Specifications
Energy Balance		Preliminary Layout		Definitive Estimate
Project Charter		Preliminary Schedule		Project Execution Plan
		Preliminary Estimate		Preliminary 3-D Model
				Electrical Equipment List
				Line List

* Wikipedia

7

FEL Gated Project Delivery Process

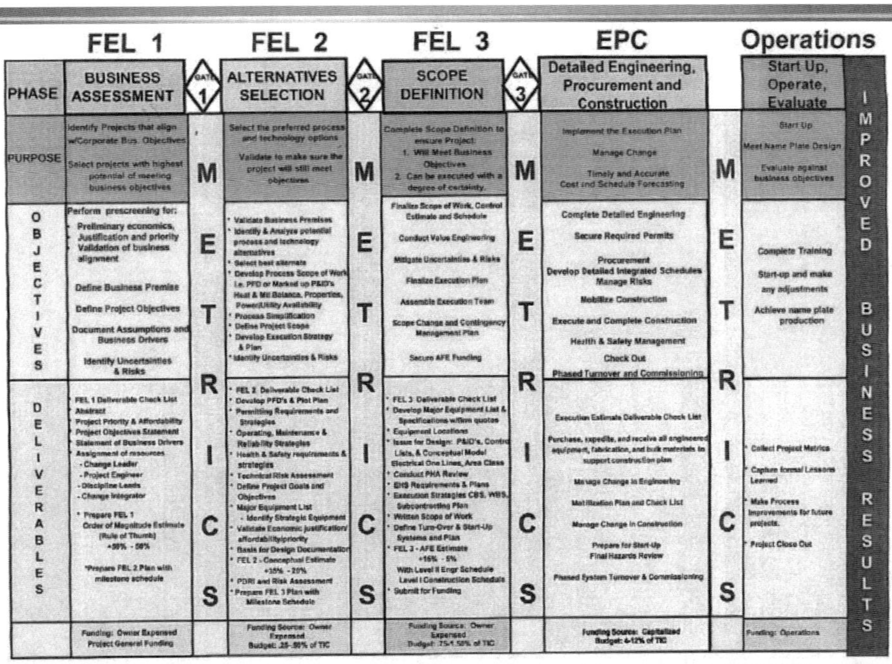

8

Example: Tanks & Interfaces

9

Critical Interfaces in Tank Design and Construction

	Tank Vendor	Agitator Vendor	Detailed Engineer	Structural / Civil Engineer	Piping Contractor	Concrete Contractor	Structural Steel Contractor	Electrical Contractor
Technology Provider	■	■	■	■				
Tank Vendor		■	■	■	■	■	■	
Agitator Vendor	■		■				■	■
Detailed Engineer	■	■		■	■	■	■	■
Structural / Civil Engineer	■		■		■	■	■	
Piping Contractor	■		■	■		■	■	■
Concrete Contractor	■		■	■	■		■	
Structural Steel Contractor	■	■	■	■	■	■		■
Electrical Contractor		■	■		■		■	

10

Division of Responsibilities

Delta-T HIDE ALL UPDATE REV

	TECH PROVIDER			CLIENT			CONSTRUCTOR		
	ENGR	PURCH	CONST	ENGR	PURCH	CONST	ENGR	PURCH	CONST
• EPC PROJECT CONTROLS									
PROJECT MANAGEMENT									
PROJECT ENGINEERING									
COST CONTROL									
ESTIMATING									
SCHEDULING									
QA/QC									
• PROCUREMENT									
PROCESS									
REQUISITIONS									
PURCHASING									
PROCESS EQUIPMENT									
MAJOR ELECTRICAL GEAR									
FIELD FABRICATION TANKS									
PRE-ENGINEERED BUILDINGS									
SUBCONTRACTS									
EXPEDITING									
INSPECTION									
SUBCONTRACTING									

11

Tools for Success

- **Mind the Interfaces**
  - Lack of good communication at interfaces between disciplines and organizations leads to errors, delays and cost overruns
- **Detailed Scope Definition**
  - Tools such a Division of Responsibilities are necessary in managing interfaces
- **Dedicated Project Manager**
  - Disciplined, highly organized, detail oriented

12

AIR PRODUCTS

Global Dimensions of New Product Development: A Case Study

AIChE 2012 Process Development Symposium
Michael Popule, Air Products

Presentation Outline

- Project Background
- Project Scope and Objectives
- Team Structure
- Defined Roles
- Project Considerations
- Lessons Learned

Formation of a Network

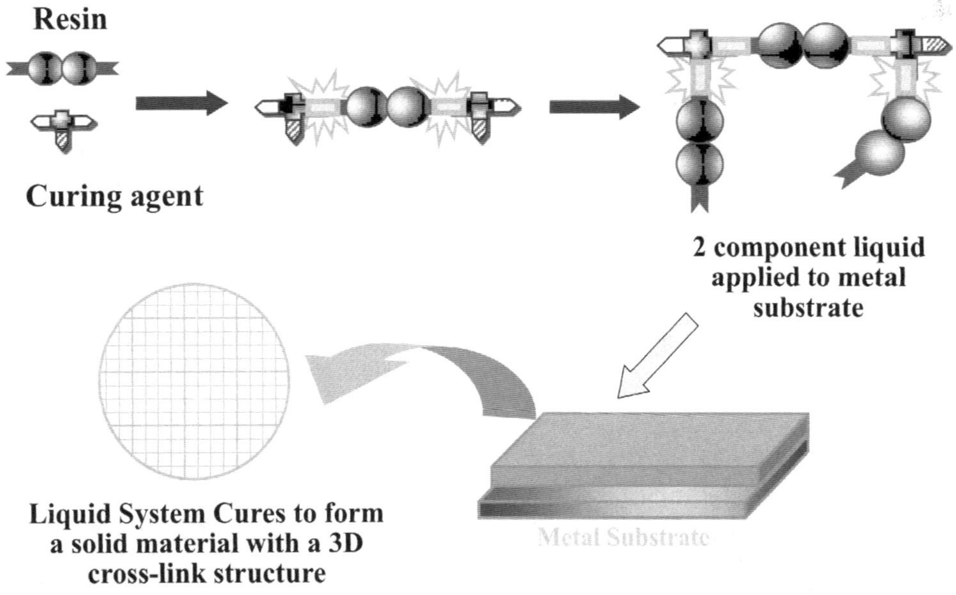

Project Scope

- Primary: To develop a new platform of products with a significantly lower cost base compared to current offerings and provide greater security of supply to customers in all regions.
- Secondary: Identify unique performance attributes that enable growth into new market segments and adjacent markets.

AIR PRODUCTS

A Versatile New Platform

New Platform of Products

AIR PRODUCTS

Project Objectives

- Develop consistent manufacturing processes for the new platform of products to enable scale-up in multiple assets at reduced manufacturing cost compared to standard
- Ensure IP is filed and protected in all world regions
- Ensure regulatory registration in all world regions
- Establish supply chain for all required raw materials and finished product distribution
- Validate performance attributes in multiple market segments with global customers
- Develop a commercial introduction plan
- Meet or exceed financial targets for the project

7

AIR PRODUCTS

A multi-disciplinary team works in parallel to . . .

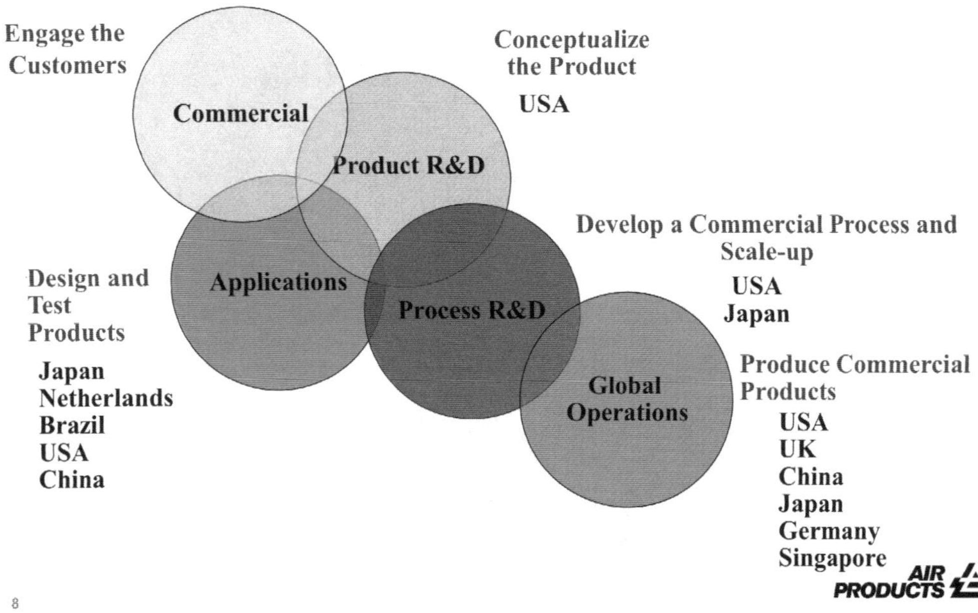

8

AIR PRODUCTS

Global Business with Multiple Centers of Excellence

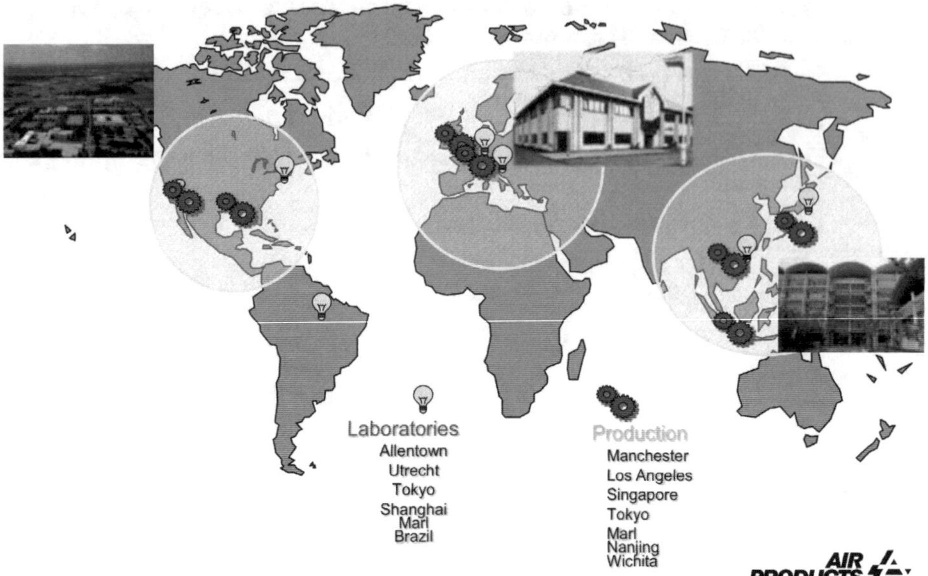

AIR PRODUCTS

Team Structure

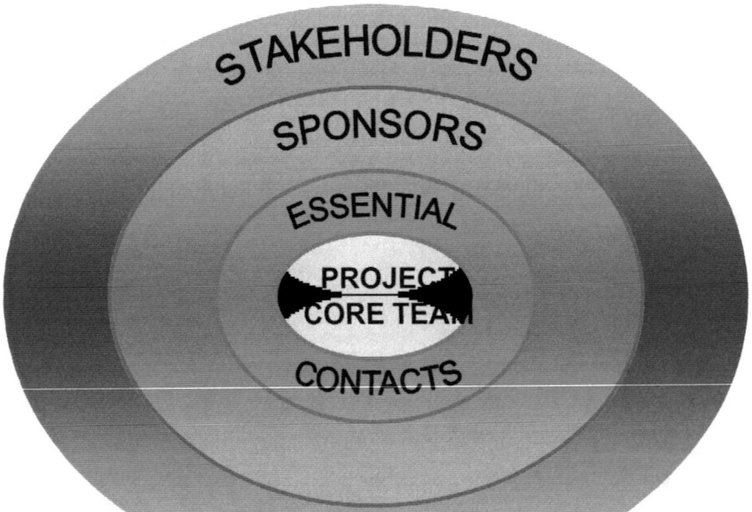

AIR PRODUCTS

Project Core Team

- Consists of a 'small' group of cross-functional members
- Represent multiple regions to invigorate innovation process and accelerate knowledge distribution. Think globally, act locally.
- Have the time to actively engage in the project.
- Are passionate about the opportunity to contribute particularly with global impact.
- Are optimistic about the probability of success.
- Take ownership of the project.
- Support the Project Manager.

Essential Contacts / Associate Members

- Are essential to engage in order for the project to succeed.
- Represent a critical function or service that will supply an important contribution sometime during the life of the project.
- Need to be proactively approached and engaged at the appropriate time.
- May transfer in and out of the 'core team' for some duration of the project.
- Have the time to contribute as required.
- Become advocates for the project and will support direction put forward by the Core Team.

Sponsors / Steering Team

The term 'Sponsor' refers to the person or team of people who:

- Providing support to the Core Team for the success of the project
- Prioritize projects within the portfolio
- Delegate authority and set boundaries
- Establish project expectations
- Provide resources
- Evaluate performance
- Champion the effort
- Handle Top Management interactions

13 Property of Air Products and Chemicals, Inc.

Innovate Project Manager

The term 'Project Manager' refers to the person who is responsible for ensuring that:

- Sponsorship is well established
- A suitable project team is assembled
- The scope, schedule and budget are clarified
- The appropriate work processes, tools and safety efforts are in compliance
- The project is executed in an efficient manner
  - Keep Project Plan updated
  - Gather Core Team input to Project Plan
  - Ensure Documentation is updated
- Stakeholder communications occur as necessary

14 Property of Air Products and Chemicals, Inc.

Roles Defined in New Product Introduction

- Overall Project Mgr.
- NP Execution Lead
- Marketing
- Product Research
- Applied Technology
- Process R&D
- GO
- Supply Chain
- Distribution & Logistics
- Asset Management
- Plant Process Eng
- EH&S
- Purchasing
- Customer Service
- System Data Technicians

15 Property of Air Products and Chemicals, Inc.

AIR PRODUCTS

Technical/Commercial Hurdles

- Working with various reactor designs across asset base (mixing, heat transfer)- evaluate process sensitivity and key control parameters
- Raw materials in each region have specification differences- impact on chemistry
- Process safety impacted by asset selection
- Regulatory rules are not consistent (e.g. polymer definition) and rules are changing
- Different market segments require different performance attributes and product characteristics (e.g. labeling, cost in use)
- Regional differences in how product is used and what is acceptable to customers
- Customer intimacy varies between multinationals and local customers
- Establish points of contact with customer

16 Property of Air Products and Chemicals, Inc.

AIR PRODUCTS

Team Execution Strategy

- Work in phases and don't do everything at once
- Clearly document discussions, decisions and actions.
- Use both synchronous and asynchronous communication
- Coordinate alignment and communications outside the core team (single voice)
- Segment what is global and what is regional
- Identify critical support members outside the core team

AIR PRODUCTS

Lessons Learned

- Need a full time dedicated project manager for broad global projects
- Strong, highly experienced core team is needed to handle complexity
- Team leader needs to take an active role in "on-boarding" new team members
- Leveraging a diverse set of resources requires spending time on alignment. Take time to set up the norms/ground rules (including decision making, communications used, ...) during the first meeting and review them during the following virtual meetings
- Take time to build trust and relationships prior to startup. An initial face to face meeting is best but using pictures and storytelling also works
- Need to back away from assumptions about how things would get done and communicate. Take time to frame the right questions

AIR PRODUCTS

Thank you

tell me more

www.airproducts.com

Transfer of Pharmaceutical Finished Product Processes to Third Party Manufacturers

Peragine, J. F., Bristol-Myers Squibb Company
Originally presented on: 6/7/2012 11:00:00 - 11:30:00

Pharmaceutical companies routinely use third party manufacturers (TPMs) to manufacture their drug products, in addition to their own manufacturing sites. For currently marketed products, the technology transfer and implementation of the commercial process can be especially challenging since the equivalence of the manufacturing processes at the current and new sites needs to be demonstrated. The manufacturing process at the new site needs to yield final drug product which has the same quality attributes as the product from the current site. Process/product equivalence is also important to demonstrate for the required FDA and multinational registrations before the new site can be approved for manufacturing.

Examples of some challenges of technology transfer of pharmaceutical finished products to TPMs are given in case studies for three different pharmaceutical dosage forms - capsules, sterile liquid, and tablets. The capsule product was manufactured at two different in-house sites and transferred to a new TPM. The sterile liquid product was manufactured at the same TPM for 40 years and was transferred to a new TPM. The tablet product is still being manufactured in-house but is being considered for transfer. For this active project, there are currently six different TPMs under consideration due to the nature of the product and complexity of the manufacturing process (so only limited information can be shared at this time).

OUTSOURCING OF MANUFACTURING SERVICES FROM THE TOLLING COMPANY'S PERSPECTIVE

Presentation by Joan Roller, Business Development Manager

CRI

Chemical, Colorant, & Contract Packaging Tolling Services
www.CRItoll.com

Chemical Industry 2012

- Over 2,200 Mergers & Acquisitions occurred between 1987 to 2009.
- Many US Chemical Site Closings.
- Outsourcing is a key element in the Chemical Industry.

Chemical Services Outsourced

- Engineering
- Synthesis
- Testing
- Drying
- Repackaging
- Blending
- Change in particle size
- Purification

Why Do Companies Outsource?

- Plant closings
- Don't have the equipment needed
- Don't want to increase the capital base
- Timing
- Capital may be less
- Cost/Value may be improved
- Product life cycle
- Bridging strategy
- Need for additional capacity
- Geographic reasons
- Environmental permitting

US Tolling Companies

- Generally smaller companies < $1 billion or separate business unit.
- Not as much market focus more "tool kit" focused.
- Many privately owned.
- Product life cycle is 3 to 8 years.
- Secrecy agreements.

Process for Handling Inquiries

Phase 1 - Information Gathering

- Time not much money
- Say "no" quickly
- Create a cost estimate
- Preliminary pricing

Phase 2 - Detail Planning

- Detail technical package
- Lab work
- Engineering work
- Capital with letter of intent or contract
- Trial

Phase 3 - Commercialization

- Contract
- Production
- Follow up

Pricing

- Toller has little information about the needed price.
- List your assumptions.
- Clear communication about what is included in the price, freight, raw materials, packaging materials.
- Logistics.
- Get the units right.
- Success rate is about 5% to 10%.

Toller's Expectation

- A good technical package.
- Accurate business details.
- Technical meetings to discuss issues.
- Site visits both to Customer's & Toller's Plant.
- Approved vendor lists of raw materials.
- Analytical test methods.

How Customer's Outsource Projects

- Organization
- Project/Technical Manager
- Purchasing/Sourcing Manager
- Regular meetings with minutes.
- Pull in technical help as necessary.
- Be involved with PHA's if applicable.
- Be involved with the start up.

Capital Requirements

- Maybe none up to $10's million.
- Capital must be recovered in the price or in special payments.
- Dealing with smaller companies the payback may need to be over a short time line.
- The better the technology transfer, the few over runs.

Build Relationships

- Customer and Toller must trust each other.
- Customer and Toller must have a good working relationship.
- Customer and Toller must keep each other well informed.
- Customer and Toller must be prepared to make adjustments to commercial agreements.

Desires on Both Sides

- Every Customer wants to make outsourcing process as easy as possible.
- Every Toller wants repeat business from a Customer.
- If business is not awarded to a Toller, Customer should be prepared to explain why.

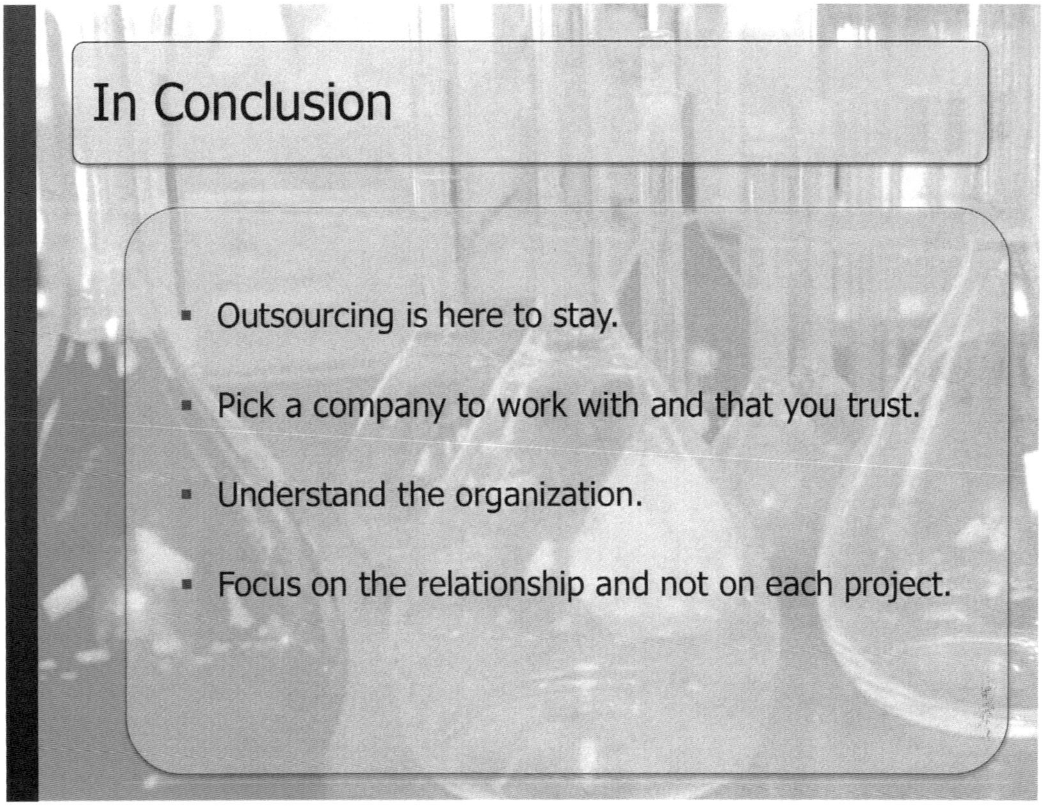
In Conclusion
Outsourcing is here to stay.
Pick a company to work with and that you trust.
Understand the organization.
Focus on the relationship and not on each project.

"Technology Transfer – Managing the Handoff"

AIChE Process Development Conference
June 5-7, 2012,
King of Prussia, PA

Ron Leng
Dow Chemical

June 2012 Ron Leng -- Dow Chemical 1

What we want to avoid

Lack of good communication between functions can lead to ineffective transfer of technology

June 2012 Ron Leng -- Dow Chemical 2

Technology Evolution

June 2012 Ron Leng -- Dow Chemical 3

4 Keys to Successful Projects

- Consistent R&D process w/ stage-gate reviews by multi-functional subject matter experts (SME)
- ***Early engagement by Manufacturing & Engineering (M&E)*** in R&D Projects
- Active ***R&D engagement*** w/ capital project and start-up teams
- Robust risk identification and management plan

  - Chemist / Engineer partners
  - Upstream & downstream cross-functional collaboration
  - Good communication and effective handoffs are important to all aspects of a project

June 2012 Ron Leng -- Dow Chemical 4

Types of Handoffs

- Chemist to Engineer
  - Process scale up from bench to pilot plant
- Research to Process Engineering
  - Capital Project --- Plant Design
- Research to Operations
  - Transfer of "know-how" to manufacturing
- Tech Transfer to a Toll Manufacturer
- Tech Transfer from a Licensor

June 2012 Ron Leng -- Dow Chemical 5

Chemist to Engineer

Goal: Translate the bench chemistry into an economical and robust process

- The bench chemists often don't fully understand the effects of their decisions on the capital for the plant & the cost of manufacturing
- Chemists tend to focus more on the reaction step, which is quite appropriate, however...
  - Rxn step may only be a small fraction of the total capital footprint of the plant
- Engineers like balanced equations with byproducts, whereas chemists often just list the molecules of interest on the chemistry diagram
- Engineers use block flow diagrams that indicate unit operations and process equipment vs. procedural activities like mix, post react, cool
- Chemists list volumes and equivalents of solvents used, whereas the engineers prefer to speak in terms of mass

June 2012 Ron Leng -- Dow Chemical 6

Chemist to Engineer

Goal: Translate the bench chemistry into an economical and robust process

- Chemists often use ***qualitative terms*** like "the mixture was thick" "work-up was tedious", "reaction proceeded smoothly", "slow filtration"
- Engineers prefer ***quantitative descriptors***, i.e. the filtration rate was 5.5 gallons / min / ft2 at a 4 inch cake depth with 45 psi applied
- Chemists often don't understand distillation
  - Use a vigreux column w/o a reflux head
  - Assume that azeotropes always have fixed ratio
- Chemists often make choices based on the short term/convenience
  - Exotic reagents, solvents, purification techniques
- Chemists don't always appreciate the impact of scale
  - Time to add a reagent, complete a distillation, or power to mix a reaction

June 2012 Ron Leng -- Dow Chemical 7

Understanding Perspective

- The R&D and M&E cultures are different

  - In R&D we want change – an unexpected result is welcome and leads to new discovery
    - We want the freedom to be creative
  - In M&E (Manufacturing & Engineering) we really don't want change, consistency and predictability are desired
    - We need a disciplined approach to getting things done

- Divergent and Convergent thinking are both good, just not at the same time.
- Need to have common "scorecard" elements

June 2012 Ron Leng -- Dow Chemical 8

Research to Manufacturing

Goal: Transmit technology and process know-how to process engineers to design, build, and operate the plant

- R&D likes to keep options open
- M&E needs convergence to move forward
- M&E think R&D can answer all questions w/ experiments
- R&D can feel that M&E are out to "break the process" by compromising design to reduce cost
- M&E need to be much more concerned w/ EH&S issues as the process moves to the plant

Bringing the Functions Together

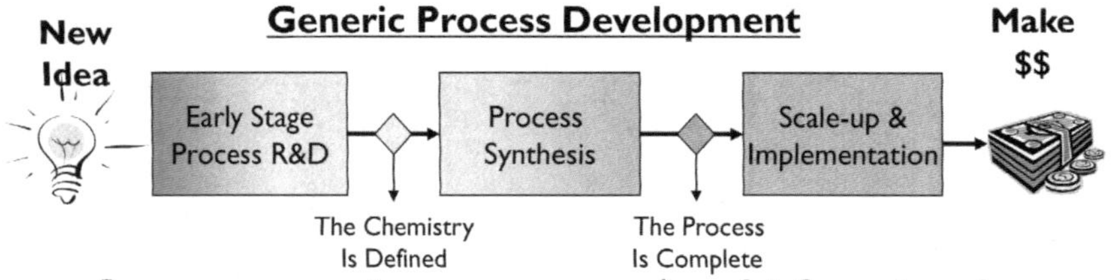

- Stage gate review process w/ multi-functional assessment
- Assign manufacturing representative on R&D led projects
- Assign R&D liaison role on capital project
- Structured, periodic reviews w/ multi-discipline engagement
  - Early-Stage EH&S review
  - Technology Status Analysis
  - Technology Selection Workshop
  - Process Technology Risk Assessment

Tech Transfer to Operations Ideas

- Assign Engineer / Chemist partners
- Locate the chemists, process development engineers, and the project team (process engineers in the same building)
- Videotape the process in the lab
- Bring the Engineers and Operators into the Laboratory – hands on process learning
- "Chemist in a Box" – plant trouble shooting guide
- Proficiency Test the operations team to demonstrate comprehension
- Jointly develop a common "score-card" for the technology implementation
- Assign chemists and development engineers roles on the start-up team

"Success Parameters"

an element of a common scorecard

Reaction Two	Units	Needs Work	Excellent	Actual Value
Catalyst Loading	mole %	> 5	< 3	1.8
Reaction Time	hours	> 16	< 8	6-8
Reaction Two Conversion	%	< 95	97-98	96.5-97
Amount of by-product X	%	> 5	< 2	< 2
Catalyst Filtration	hours	> 4	< 2	> 12
Water to Solvent ratio for Crystallization	ratio	+/- 2.5%	1.85	
Yield loss to filtrate/wash	%	> 8	< 4	4
Product Filtration Time/Wheel	hr	> 1	< 0.5	45-60
Final Purity of Drycake	%	< 80	> 85	> 86
Overall Isolated Yield	%	< 60	> 80	78

Tech Transfer to Toll Manufacturer

Goal: Transmit technology to a third party manufacturer (3PM) who will run your process

- 3PM used for particular expertise or to avoid capital
- Cultural, geographical, language, time zone challenges
- 3PM often doesn't have time or capability to do process development
- Varying level of technical competence,
  - Chemistry and engineering support
  - Sophistication of control automation
  - Adherence to EH&S standards
- Understand what level of support is needed, permitted during start-up
- Need to develop good relationship through face to face meetings
  - May be challenging due to geographical differences
- Potential for loss of intellectual property

Bid Solicitation Package

- Only provide critical info to secure a bid
- Approach it as though you were preparing the bid – eliminate extraneous info
  - Necessary to fit the process into existing assets
    - Temps, pressure, materials of construction
  - Predict throughput / cycle time (reaction times, assumptions for separations, etc.)
  - Understand EH&S issues (heats of rxn, flash points, key IH data)
  - Usually don't need to know the exact chemistry or the intricacies of the technology at this stage
- Once contractor is selected, follow up w/ detailed technology package and face to face meetings

Manufacturing Option Rating

Selection Criteria	Capital	Conversion cost @ maturity $/kg	Yield	Product Integrity	Technical Resources	Environmental Capability	Recycle Equip is Available	Reliability of Supply	EH&S	Ownership of Improvements	Technology Protection	Contractor Security Size of Company	Total
Weighting Factor	6	9	6	9	3	3	3	9	3	1	1	3	
Manufacturer													
Internal Option A	1	6	6	6	9	9	1	6	9	9	9	6	324
External Option 1	9	1	3	1	6	3	6	3	6	3	3	6	204
External Option 2	6	3	3	6	6	6	6	3	3	3	3	3	240
External Option 3	9	9	3	6	3	3	9	3	3	3	3	1	297

- Useful to define what is necessary for success
- Useful for communicating decision to stakeholders

Technology In-license

- Know your licensor's experience –
  - for some it is their business model, others first time
  - Provide a list of specific questions to go w/ the list of technical documents
- Language barriers (Korea, Japan, China, etc.)
  - Recruit a native speaker who knows the language, culture, and technology
  - Do your homework on cultural norms, "reach across the table" (greeting, gifts, socializing, etc)
- Geographical barriers
  - It may be difficult and costly, but face to face meetings are essential

Conclusion

- Understanding the perspectives of who you are receiving or transferring technology to is the key to managing the hand-off successfully!
- Expertise, Experience, Function, Culture differences all need to be considered
- You have to invest the time to "shake hands at the elbow"
- Thank-you!

AUTHOR INDEX